Grade 1

Discovery Education | SCIENCE TECHBOOK

Unit 3
Shadows, Light, Motion in the Sky

Discovery EDUCATION™

Copyright © 2020 by Discovery Education, Inc. All rights reserved. No part of this work may be reproduced, distributed, or transmitted in any form or by any means, or stored in a retrieval or database system, without the prior written permission of Discovery Education, Inc.

NGSS is a registered trademark of Achieve. Neither Achieve nor the lead states and partners that developed the Next Generation Science Standards were involved in the production of this product, and do not endorse it.

To obtain permission(s) or for inquiries, submit a request to:
Discovery Education, Inc.
4350 Congress Street, Suite 700
Charlotte, NC 28209
800-323-9084
Education_Info@DiscoveryEd.com

ISBN 13: 978-1-68220-789-5

Printed in the United States of America.

6 7 8 9 10 CWM 26 25 24 23   B

© Discovery Education | www.discoveryeducation.com

Acknowledgments

Acknowledgment is given to photographers, artists, and agents for permission to feature their copyrighted material.

Cover and inside cover art: Pixabay

Table of Contents

Unit 3

Letter to the Parent/Guardian.................................... v

Shadows, Light, Motion in the Sky viii

   Get Started: Making a Shadow Theater................ 2

Unit Project Preview: Experimenting with Shadows ... 4

Concept 3.1

Light Sources.. 6

   Wonder 8

      Let's Investigate Cave Diving 10

   Learn 18

   Share 48

Concept 3.2

Light in Action 58

   Wonder 60

      Let's Investigate Shadows........................... 62

   Learn 70

   Share 100

Concept 3.3
Sunlight 108
- **Wonder** 110
  - Let's Investigate Sun and Moon 112
- **Learn** 118
- **Share** 150

Concept 3.4
Moon on the Move 158
- **Wonder** 160
  - Let's Investigate the Night Sky 162
- **Learn** 168
- **Share** 188

Unit Project
Unit Project: Experimenting with Shadows 198

Grade 1 Resources
Bubble Map R1
Safety in the Science Classroom R3
Vocabulary Flash Cards R7
Glossary R17
Index R29

Dear Parent/Guardian,

This year, your student will be using Science Techbook™, a comprehensive science program developed by the educators and designers at Discovery Education and written to the Next Generation Science Standards (NGSS). The NGSS expect students to act and think like scientists and engineers, to ask questions about the world around them, and to solve real-world problems through the application of critical thinking across the domains of science (Life Science, Earth and Space Science, Physical Science).

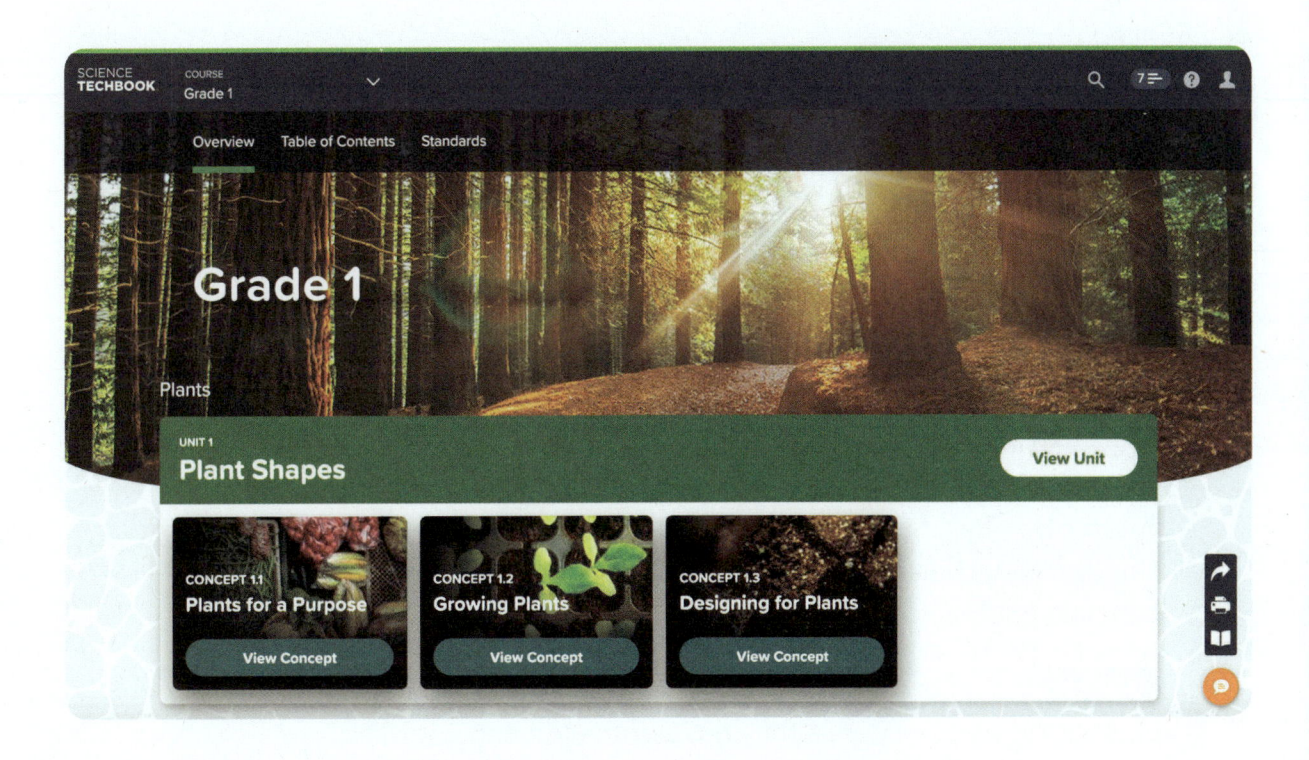

Science Techbook is an innovative program that helps your student master key scientific concepts. Students engage with interactive science materials to analyze and interpret data, think critically, solve problems, and make connections across science disciplines. Science Techbook includes dynamic content, videos, digital tools, Hands-On Activities and labs, and gamelike activities that inspire and motivate scientific learning and curiosity.

You and your child can access the resource by signing in to www.discoveryeducation.com. You can view your child's progress in the course by selecting Assignments.

Science Techbook is divided into units, and each unit is divided into concepts. Each concept has three sections: Wonder, Learn, and Share.

Units and Concepts Students begin to consider the connections across fields of science to understand, analyze, and describe real-world phenomena.

Wonder Students activate their prior knowledge of a concept's essential ideas and begin making connections to a real-world phenomenon and the **Can You Explain?** question.

Learn Students dive deeper into how real-world science phenomenon works through critical reading of the Core Interactive Text. Students also build their learning through Hands-On Activities and interactives focused on the learning goals.

Share Students share their learning with their teacher and classmates using evidence they have gathered and analyzed during Learn. Students connect their learning with STEM careers and problem-solving skills.

Within this Student Edition, you'll find QR codes and quick codes that take you and your student to a corresponding section of Science Techbook online. To use the QR codes, you'll need to download a free QR reader. Readers are available for phones, tablets, laptops, desktops, and other devices. Most use the device's camera, but there are some that scan documents that are on your screen.

For resources in Science Techbook, you'll need to sign in with your student's username and password the first time you access a QR code. After that, you won't need to sign in again, unless you log out or remain inactive for too long.

We encourage you to support your student in using the print and online interactive materials in Science Techbook on any device. Together, may you and your student enjoy a fantastic year of science!

Sincerely,

The Discovery Education Science Team

Unit 3: Shadows, Light, Motion in the Sky | vii

Discovery
EDUCATION

Unit 3
Shadows, Light, Motion in the Sky

Get Started

Making a Shadow Theater

Shadow puppets can be used to tell stories. At the end of the unit, you will be able to use the information you learned to create a shadow theater and performance.

Watch a video about how puppeteers make different shapes with shadows.

Quick Code: us1505s

Making a Shadow Theater

Think About It

Look at the picture. **Think** about the following questions:

- What causes shadows?

- What happens when there is no light?

- What objects are in the sky and how do they seem to move?

- When will the sun set tomorrow?

- How does the moon's appearance change over each month?

Shadows

Unit 3: Shadows, Light, Motion in the Sky | 3

Unit Project Preview

Design Solutions Like a Scientist

Hands-On Engineering: Experimenting with Shadows

In this activity, you will decide what source of light to use in a shadow puppet show.

Quick Code: us1506s

Shadow Puppet Show

- **SEP** Asking Questions and Defining Problems
- **SEP** Constructing Explanations and Designing Solutions
- **SEP** Planning and Carrying Out Investigations
- **SEP** Analyzing and Interpreting Data
- **CCC** Patterns
- **CCC** Cause and Effect
- **CCC** Structure and Function

Ask Questions About the Problem

You are going to design a shadow puppet show that can be performed during the day and at night. **Write** some questions you can ask to learn more about the problem. As you work on activities throughout the unit, **write** down answers to your questions.

CONCEPT
3.1

Light Sources

Student Objectives

By the end of this lesson:

- ☐ I can investigate how objects can be seen when light hits them and share my findings.

- ☐ I can investigate how objects can be seen when the objects give off their own light and share my findings.

Key Vocabulary

- ☐ reflect
- ☐ source
- ☐ star
- ☐ sun
- ☐ technology

Quick Code: us1508s

Activity 1
Can You Explain?

Meena

How does light help us see objects?

we cant see in the drak But when the light is of we cant see.

Quick Code: us1509s

3.1 Light Sources | **9**

3.1 | Wonder How does light help us see objects?

Activity 2
Ask Questions Like a Scientist

Quick Code: us1510s

Cave Diving Meena

Watch the video. **Answer** the questions.

Let's Investigate Cave Diving

What things in the video gave off light?

flashlight

Can you think of times when you tried to see things in a dark room?

can't see

How was that similar to the divers in the video? How was it different?

Same - need light to see

What questions do you have about cave diving?

Your Questions

when it is brke you cant see?

3.1 Light Sources | 11

Activity 3
Analyze Like a Scientist

Quick Code: us1511s

Exploring Caves

Read about exploring caves. Then, **complete** the activity.

> Read Together

Exploring Caves

There are many places on Earth that humans have not explored. New types of animals and plants are often found in unexplored areas.

Underwater caves are hard for humans to explore. Explorers need tools to help them see. One tool explorers use is a flashlight.

Exploring Underwater Caves

Write or **draw** a description of a time you went exploring.

3.1 Light Sources | 13

3.1 | Wonder How does light help us see objects?

Activity 4
Observe Like a Scientist

Quick Code: us1512s

Exploring Outer Space

Watch the video. Then, **talk** about the question.

Video

Exploring Outer Space

Talk Together

Now talk about telescopes that we use to explore outer space. How do telescopes use light to help us learn about objects in space?

SEP Asking Questions and Defining Problems

Activity 5
Observe Like a Scientist

Quick Code: us1513s

Palau's Underwater Caves

Watch the video. Then, **talk** about the question.

Palau's Underwater Caves

Talk Together

Now, talk together about underwater caves. Where does the light in the cave come from?

The light comes from the sun.

SEP Asking Questions and Defining Problems

3.1 Light Sources

3.1 | Wonder How does light help us see objects?

Activity 6
Evaluate Like a Scientist

Quick Code: us1514s

What Do You Already Know About Light Sources?

Discussing Light

Look at the picture. What do you wonder?

Lake on a Sunny Day

What Are Sources of Light?

Circle the light sources that people make.

Draw a box around light sources found in nature.

3.1 Light Sources   17

3.1 | Learn How does light help us see objects?

Where Does Light Come From?

Activity 7
Observe Like a Scientist

Quick Code: us1515s

Light Scavenger Hunt

Watch the video. Then, **complete** the activities.

Light Scavenger Hunt

Talk Together

Now, talk together about light. Why is it hard to see things in a dark closet?

SEP Constructing Explanations and Designing Solutions

Draw a picture of your classroom and circle the light sources.

EXIT light
sun
flash light

Draw a new design to add more light to your classroom. Include at least three different sources of light.

Bath room light
nitlight
lava lamp

3.1 Light Sources | 19

Activity 8
Analyze Like a Scientist

Quick Code: us1516s

How Does Light Move?

Read about light and how it moves. Then, **answer** the questions.

Read Together

How Does Light Move?

Light is a kind of energy. Light always comes from a source. Without a light source, you cannot see.

Sunlight

Light travels in a straight line starting at the light source. Some light sources make a narrow path of light. Some light sources make light go in all directions.

SEP Developing and Using Models

What kind of light source could make light go in a narrow path? **Write** or **draw** an example.

sun   flashlight ✓

What kind of light source could make light go out in all directions? **Write** or **draw** an example.

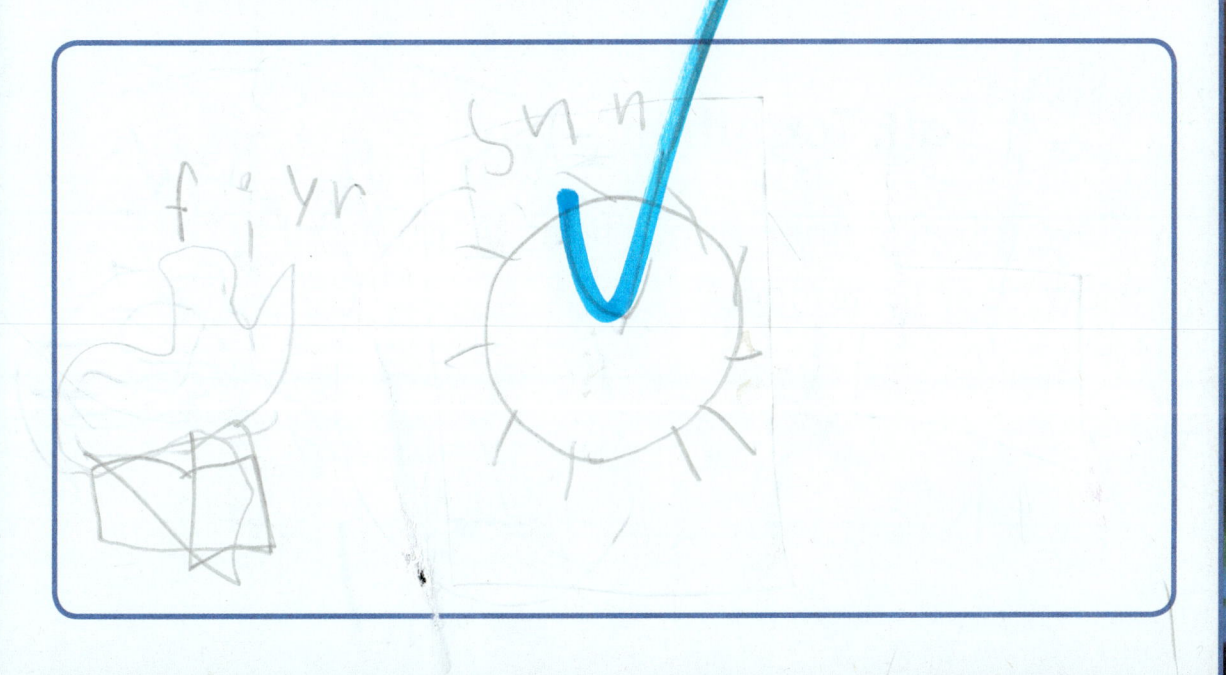

fiyr   sun ✓

3.1 Light Sources | 21

3.1 | Learn How does light help us see objects?

Activity 9
Observe Like a Scientist

Quick Code: us1517s

What Is Light?

Watch the video. Then, **talk** about the questions.

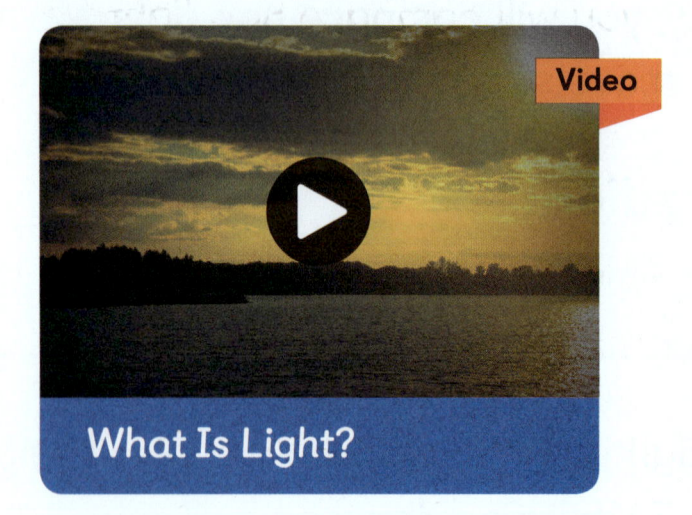

What Is Light?

Talk Together

Now, talk together about light. What is light? How does light move?

Activity 10
Investigate Like a Scientist

Quick Code: us1518s

Hands-On Investigation: The Path of Light

In this activity, you will compare how light from different light sources travels.

Make a Prediction

You are going to use different shapes of light sources. **Write** or **draw** your predictions.

How will the light bulb change how light travels?

Type of Light Bulb	My Predictions
Pointed Light Bulb	
Round Light Bulb	
One Flashlight	
Three Flashlights	

SEP Planning and Carrying Out Investigations   **CCC** Patterns

3.1 Light Sources

3.1 | Learn   How does light help us see objects?

What materials do you need?
(per group)

- Clear light bulb
- Pointed light bulb
- Clamp on light with shade
- Flashlight
- Observation chart
- Batteries, size D

What Will You Do?

Take turns switching on each type of light bulb. Point the light at the wall. **Write** or **draw** what you see in the chart.

Type of Light Bulb	Observations/Drawings
Pointed Light Bulb	
Round Light Bulb	

Next, **turn on** one flashlight and **point** it at the wall.
Write or **draw** what you see in the chart.

Then, **turn on** three flashlights. What happens?
Write or **draw** what you see in the chart.

Type of Light Bulb	Observations/Drawings
One Flashlight	
Three Flashlights	

3.1 Light Sources | 25

3.1 | Learn How does light help us see objects?

Think About the Activity

Describe the light from the different light bulbs.

What did the path of the pointed light bulb look like?

_____

_____

_____

What did the path of the round light bulb look like?

_____

_____

_____

How was the path of light from one flashlight different from the path from three flashlights?

_____

_____

_____

Activity 11
Evaluate Like a Scientist

Quick Code: us1519s

Shape of Light

Look at the pictures. **Draw** a circle around the objects that produce light in all directions.

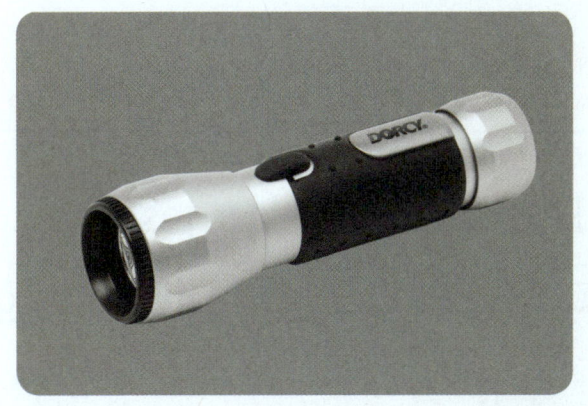

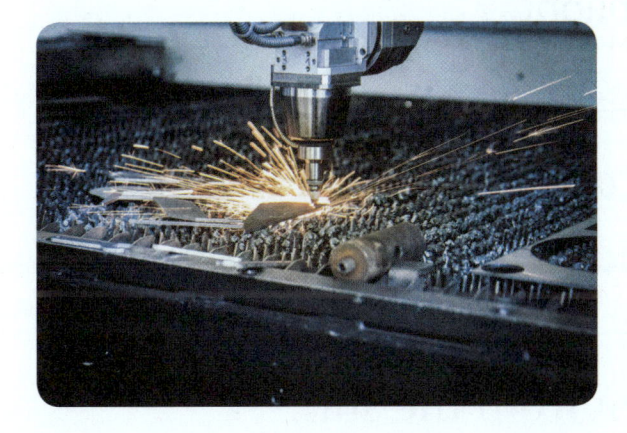

CCC Patterns

3.1 Light Sources | 27

Activity 12
Analyze Like a Scientist

Quick Code: us1520s

Sources of Light

Read the text. As you read, **underline** the sentence that tells how we see light from the moon.

Read Together

Sources of Light

The **sun** is our largest light source. Do other objects in the sky give off light?

A full moon seems to light up the night sky. If the moon gives off light, why are there times we don't see the full shape of the moon? This is because the moon is not a light source. The moon **reflects** light from the sun.

The Night Sky

28

Activity 13

Analyze Like a Scientist

Quick Code: us1521s

Stars and Light

Read the text and **underline** details about **stars** and the sun.

Read Together

Stars and Light

When you look up in the night sky, you can sometimes see many different stars.

A star is a ball of gas. Stars appear to be close to Earth, but most are very far away.

The sun is a star. Like other stars, the sun is made up of different types of gases. The sun appears to be bigger than all the other stars, but it is not. The sun is a medium-sized star. The sun looks so big because it is close to Earth.

Read Together

The sun and all other stars give off light. Because the sun is so close to Earth, it emits light in all directions.

Since other stars are farther from Earth, they don't emit light in the same way as our sun.

Stargazing

If you were to draw a picture showing how the sun emits light, you would probably draw light rays extending from the top and bottom of the sun. You would also show light rays extending from the sides of the sun. Some of these light rays travel to Earth. Light from the sun allows us to see things on Earth.

Where does Earth get its light?

Draw a picture to show how the sun gives off light.

3.1 Light Sources

Meena

Think about what you read.

How are the moon and the stars alike?
How are they different?

Write your ideas in the diagram.

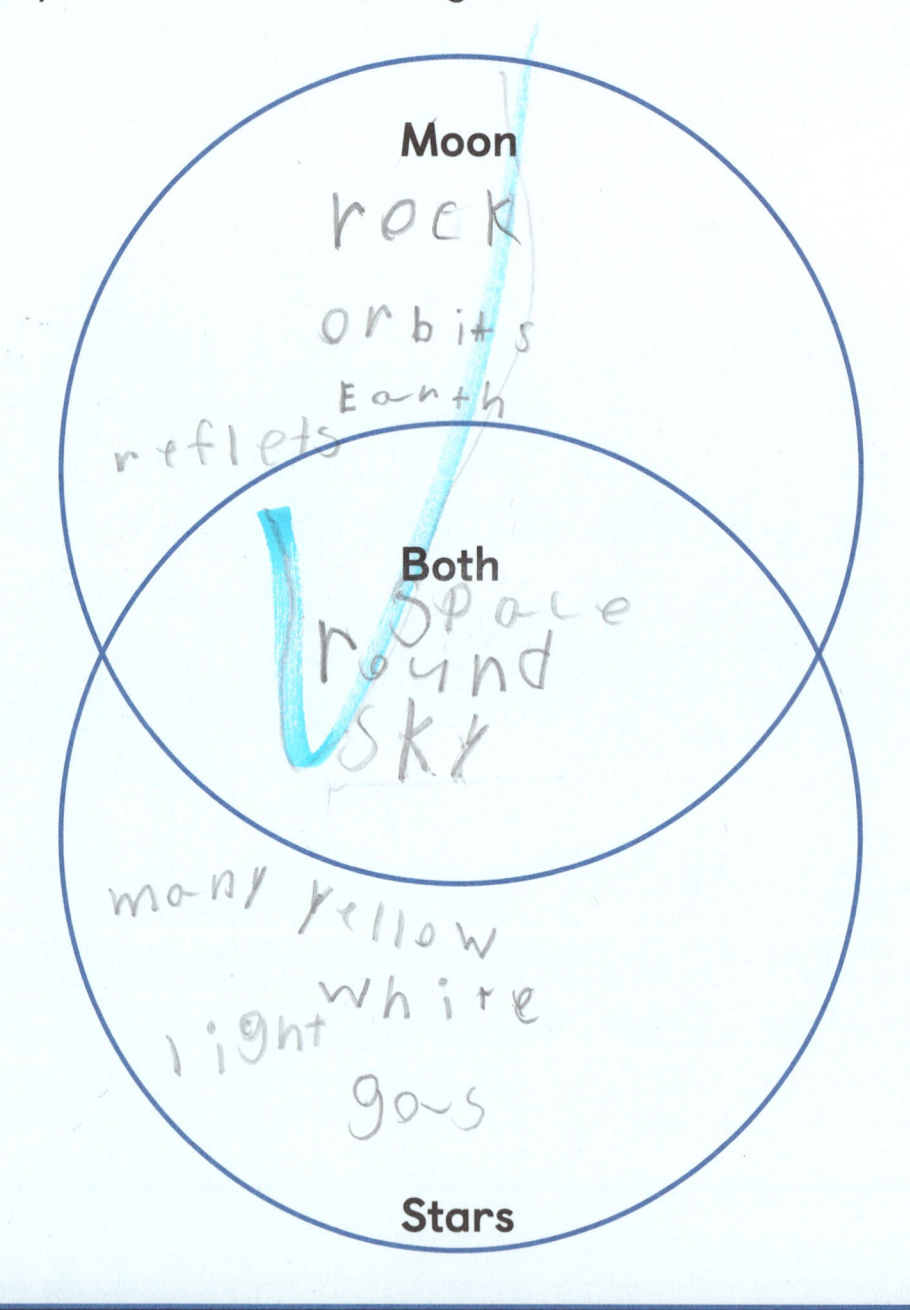

Moon
rock
orbits Earth
reflets

Both
space
round
sky

many yellow
light white
gas

Stars

Activity 14
Observe Like a Scientist

Quick Code: us1522s

Pretending to Be a Light Source

Watch the video. **Talk** with a partner about light sources. Then, **answer** the question.

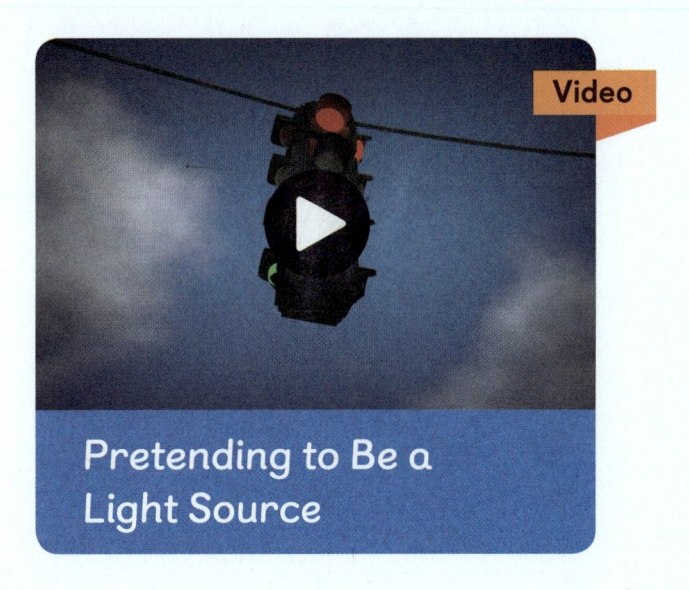

Video

Pretending to Be a Light Source

Talk Together

Now, talk about light sources. What are some different light sources?

SEP Obtaining, Evaluating, and Communicating Information

3.1 Light Sources

3.1 | Learn How does light help us see objects?

Work with a group to plan a skit about exploring a cave.

Draw a picture of your cave.

Use this chart to plan your skit.

Skit Planner	
What Tools Will We Need?	What Sights Will We See?
What Problem Might We Find?	How Can We Solve the Problem?

3.1 Light Sources | 35

How Do We Use Light?

Activity 15
Analyze Like a Scientist

Quick Code: us1523s

How Do We Use Light?
Read about using light.

Read Together

How Do We Use Light?

Think about when you turn off lights before bed at night. Without a light source, you could not see your bed, your walls, or your hand in front of your face!

To see an object, your eye has to collect light. Your eye collects the light and tells your brain the colors of the objects. Your eye has a hard time collecting light in a dark room or area.

Look at the picture, and then **answer** the question.

Meena

Seeing under the Covers

How are the brother and sister able to see in the dark?

The brother and sister able to see in the dark? Beacus the tablit has a light?

3.1 Light Sources | 37

3.1 | Learn How does light help us see objects?

Activity 16
Observe Like a Scientist

Quick Code: us1524s

How We Use Light

Watch the video. Then, **talk** about the questions.

How We Use Light (Video)

Talk Together

Now, talk together about how we use light. What groups of different lights were shown in the video?

Activity 17
Observe Like a Scientist

Quick Code: us1525s

Light at Work

Watch the videos. Then, **answer** the questions.

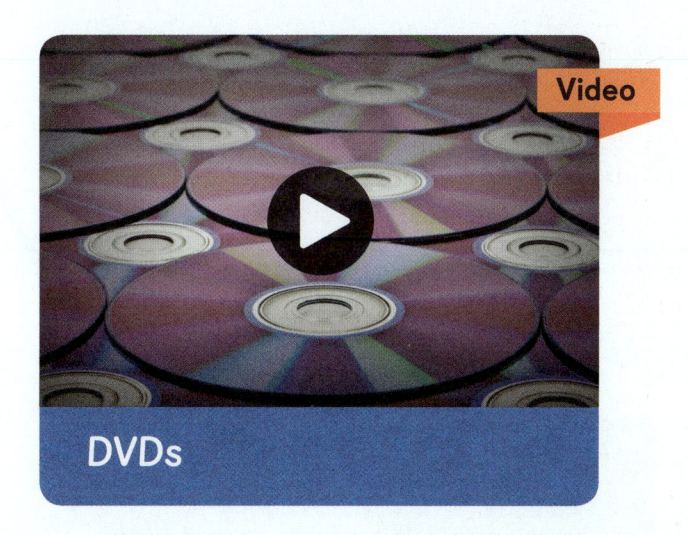

DVDs (Video)

Does the light in a DVD player travel in a narrow path, or in all directions?

_____
_____
_____

3.1 Light Sources | 39

3.1 | Learn How does light help us see objects?

Rainbow (Video)

How does light create a rainbow?

when the rain stops the light comes it crats a Rainbow

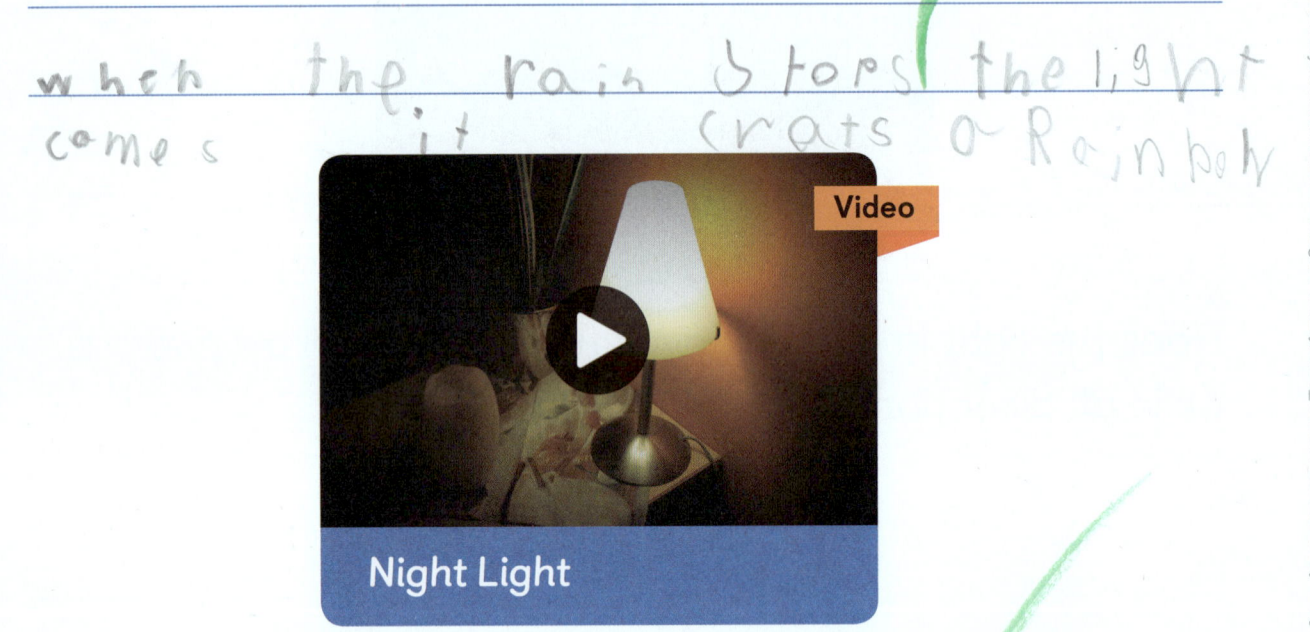

Night Light (Video)

Does the light in a night light travel in a narrow path, or in all directions?

all directions?

Activity 18

Investigate Like a Scientist

Quick Code: us1526s

Hands-On Investigation: Making a Pinhole Camera

In this activity, you will make and use a pinhole camera.

Make a Prediction

How do you think the pinhole camera will be different from a regular camera?

Write or **draw** your prediction.

> the Pin hole can t fit in own Pokit

SEP Planning and Carrying Out Investigations

CCC Cause and Effect

3.1 Light Sources | 41

3.1 | Learn How does light help us see objects?

What materials do you need?
(per group)

- Pencils
- Empty shoebox with a square cut into one side
- Metric ruler
- Wax paper
- Clear tape
- Blanket or small sheet

HANDS-ON INVESTIGATION

What Will You Do?

Tape the wax paper over the square that is cut into the shoebox. **Punch** a hole in the other end of the shoebox using the sharp pencil. **Draw** a picture of your pinhole camera!

[Student drawing showing a shoebox labeled "Pin hole" on one side and "wax paper" on the other]

Cover your head and the camera with the blanket.
Point the camera at an object.

What happened? **Write** or **draw** what you saw.

I saw a hand and the Bord and I saw a Sado.

Box

Wax

fingr

3.1 Light Sources | 43

3.1 | Learn How does light help us see objects?

Move your camera closer to the object. **Write** or **draw** what you see.

Move your camera further from the object. **Write** or **draw** what you see.

Think About the Activity

Draw what would have happened to your image if you made the hole in your camera larger.

Then, **draw** what would have happened if you made the hole smaller.

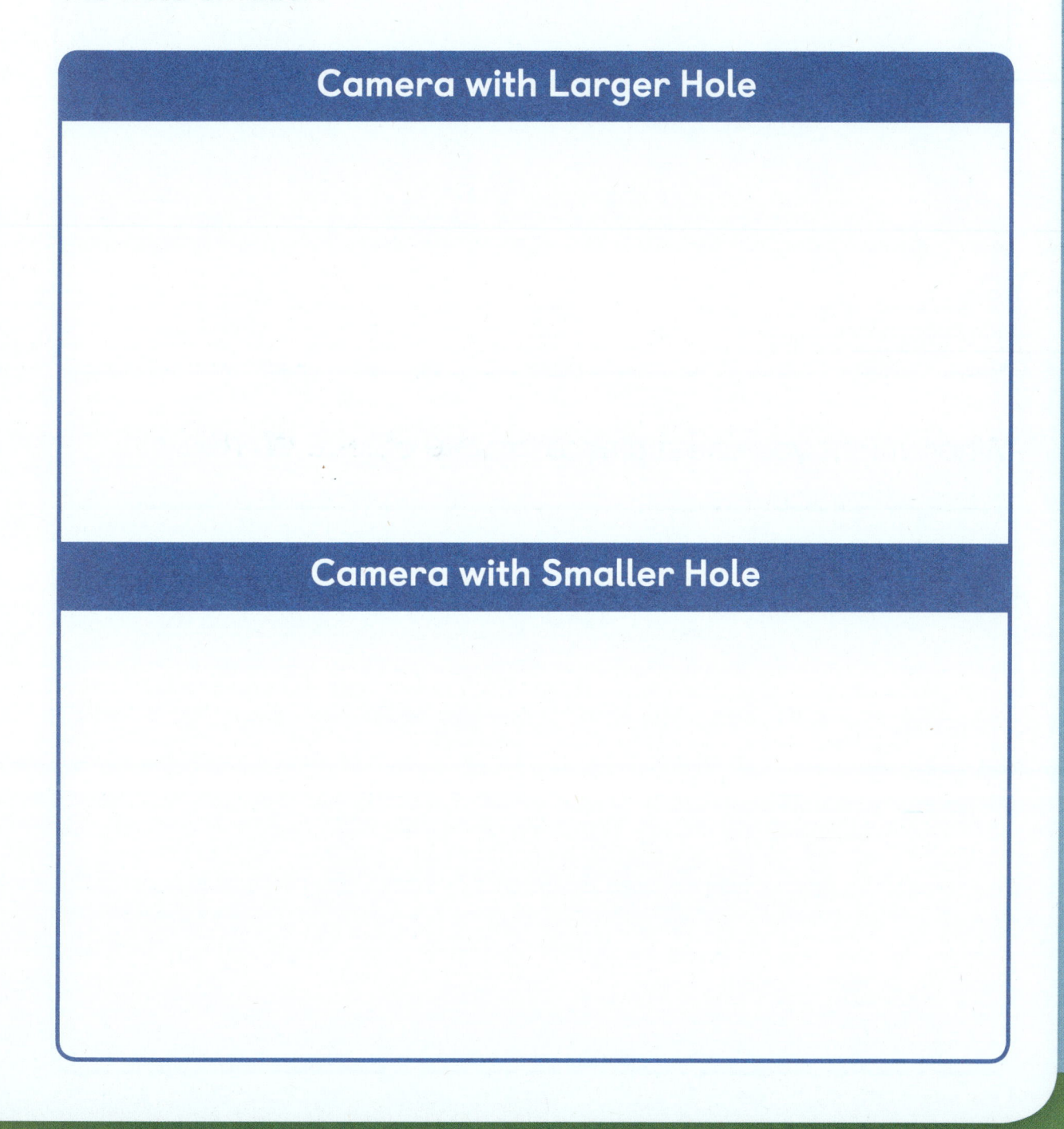

Camera with Larger Hole

Camera with Smaller Hole

3.1 | Learn How does light help us see objects?

What was the light source for your pinhole camera?

When might you use a pinhole camera?

Activity 19
Evaluate Like a Scientist

Quick Code: us1527s

Pinhole Cameras

Which picture best shows how light travels into a pinhole camera?

Circle the correct picture.

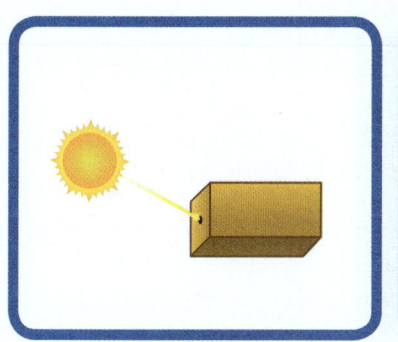

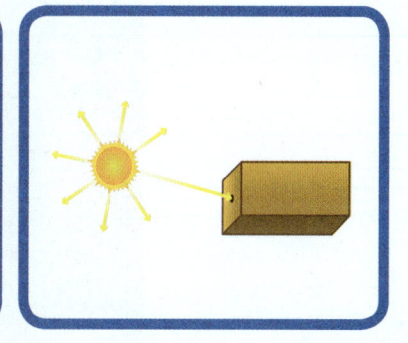

Talk Together

Now talk together. Show which picture you circled. Tell why you chose that picture.

SEP Constructing Explanations and Designing Solutions

3.1 Light Sources | 47

3.1 | Share How does light help us see objects?

Activity 20
Record Evidence Like a Scientist

Quick Code: us1528s

Seeing in a Cave

Now that you have finished learning about light sources, **look** again at the video you first saw at the beginning of the lesson.

Let's Investigate Seeing in a Cave (Video)

💬 Talk Together

How can you describe how light helps us see objects now? How is your explanation different from before?

SEP Constructing Explanations and Designing Solutions

Look at the Can You Explain? question. You first read this question at the beginning of the lesson.

> **Can You Explain?**
> How does light help us see objects?

Now, you will **use** your new ideas about the Let's Investigate Seeing in a Cave video to answer a question.

1. **Choose** a question. You can use the Can you Explain? question, or one of your own. You wrote questions at the beginning of the lesson.

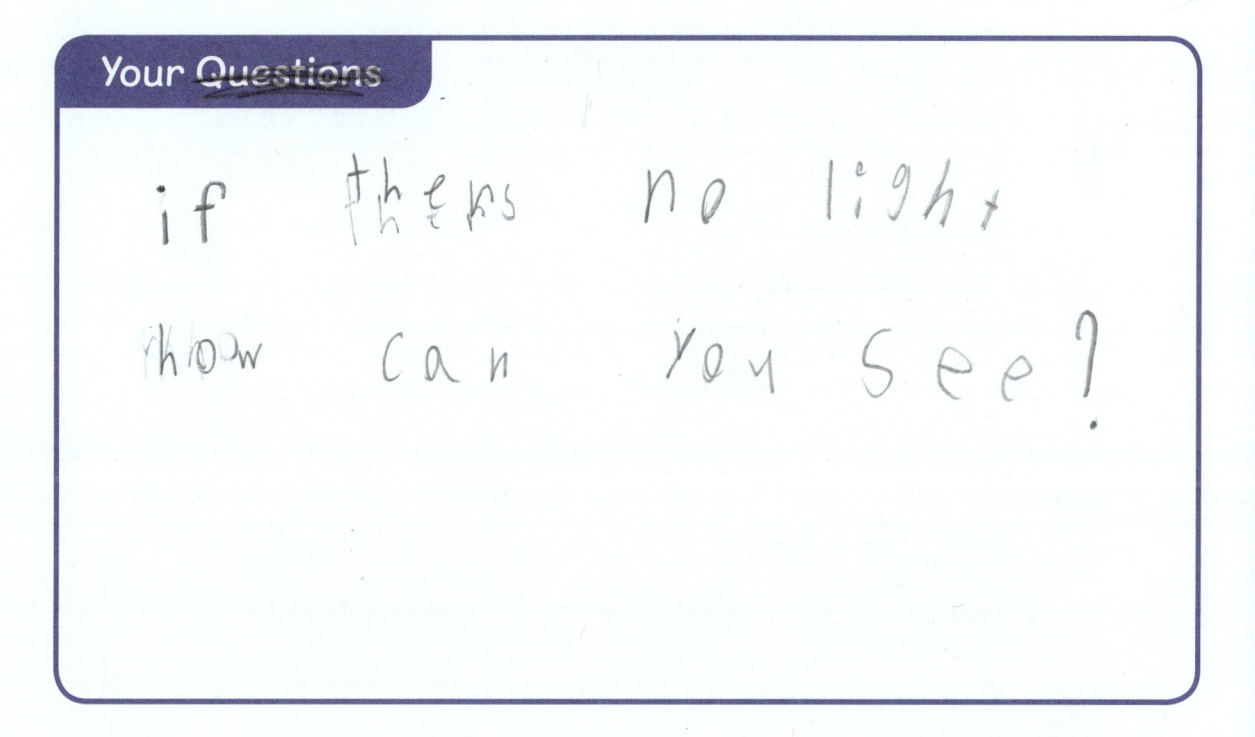

Your Questions

if thers no light how can you see?

2. **Use** the sentence starters to answer the question.

3.1 Light Sources | 49

3.1 | Share — How does light help us see objects?

Light shines from a
_____
_____
_____

Some light sources are
_____
_____
_____

Based on my observations of different light sources, light travels in different
_____
_____
_____

An example of a light source with a narrow path is
_____
_____
_____

An example of a light that travels in all directions is

We can see the sun because

We can see the moon because

The evidence I collected shows

STEM in Action

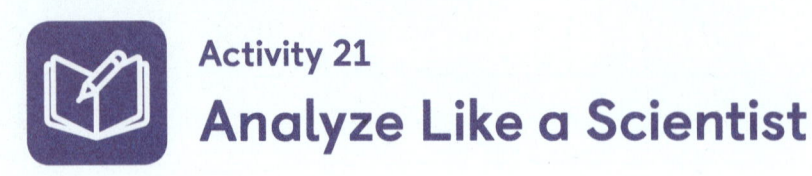

Activity 21
Analyze Like a Scientist

Quick Code: us1529s

Changing Light for a Job

Read about how an eye doctor uses light. **Circle** the sentence that tells how glasses fix color blindness.

Read Together

Changing Light for a Job

Can you imagine a job in which you use light? An eye doctor, called an optometrist, does just that.

Eye Exam

SEP Obtaining, Evaluating, and Communicating Information

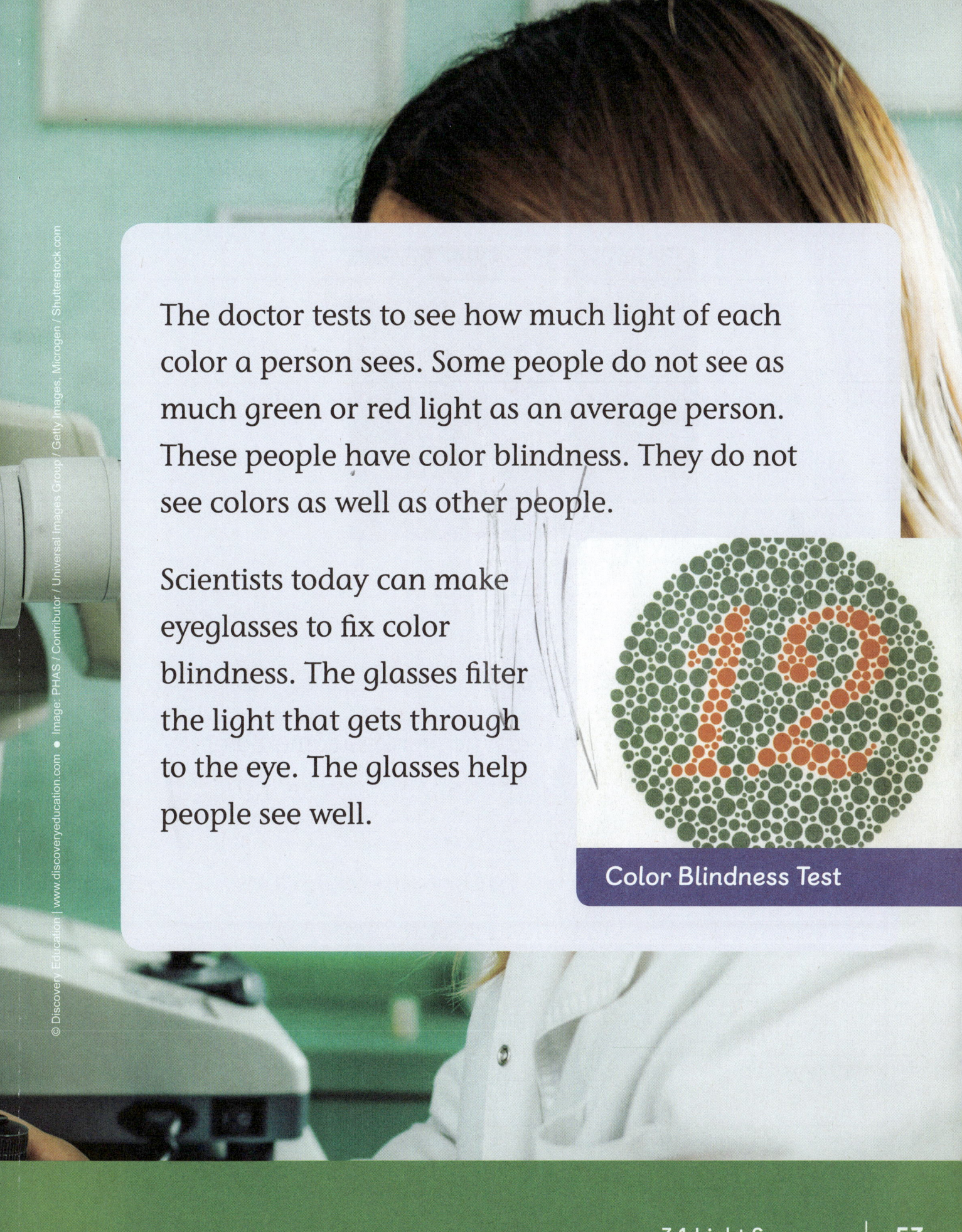

The doctor tests to see how much light of each color a person sees. Some people do not see as much green or red light as an average person. These people have color blindness. They do not see colors as well as other people.

Scientists today can make eyeglasses to fix color blindness. The glasses filter the light that gets through to the eye. The glasses help people see well.

Color Blindness Test

3.1 Light Sources | 53

Dark and Light Colors

Watch the video. Then, **talk** about the question.

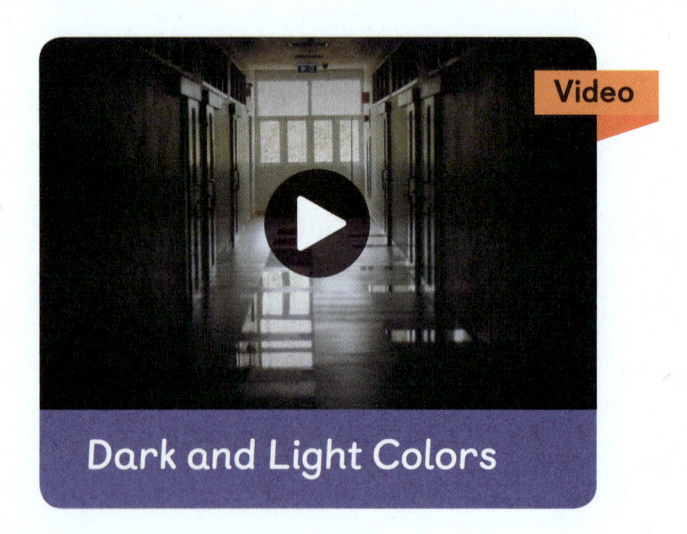

Video

Dark and Light Colors

Talk Together

Now talk with a partner. What makes some colors easier to see?

Safety First

You want to make safety vests that will show up well in low light.

Circle three vests that would be hardest to see.

3.1 Light Sources 55

3.1 | Share How does light help us see objects?

Activity 22
Evaluate Like a Scientist

Quick Code: us1530s

Review: Light Sources

Think about what you have read and seen. What did you learn? **Draw** what you have learned. Then, tell someone else about what you learned.

Talk Together

Think about what you saw in Get Started. Use your new ideas to discuss light sources.

3.1 Light Sources

CONCEPT
3.2

Light in Action

Student Objectives

By the end of this lesson:

- ☐ I can look for how light travels through transparent, translucent, and opaque materials.
- ☐ I can look for how light reflects.

Key Vocabulary
- ☐ opaque
- ☐ translucent
- ☐ transparent

Quick Code: us1532s

3.2 Light in Action

Activity 1
Can You Explain?

How does light behave when it hits an object?

mirror - reflect
solid objct - stops
transparnnt - go throgh
water - rainbow

Quick Code: us1533s

3.2 Light in Action

3.2 | Wonder — How does light behave when it hits an object?

Activity 2
Ask Questions Like a Scientist

Quick Code: us1534s

Shadows

Look at the photo. **Think** about the light and shadows.

Let's Investigate Shadows

CCC Cause and Effect

62

There is a story behind every picture. What do you notice?

When can you see shadows?

When the light comes.

What other questions do you have about light and shadows?

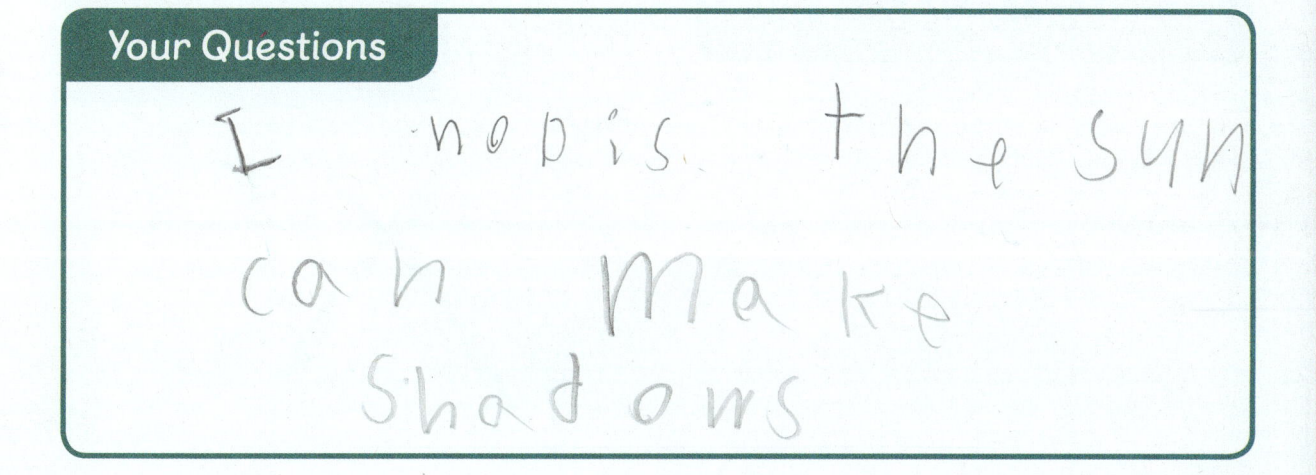

Your Questions

I nopis. the sun can make shadows

3.2 Light in Action | 63

Activity 3
Analyze Like a Scientist

Quick Code: us1535s

Shadows

Look at the picture. **Underline** the sentences that tell why you cannot see the lizard's shadow.

Read Together

Shadows

Lizards in Shade

You usually see shadows during the day. Sometimes there is not much light during the day. There is not much light at night. You do not see shadows at night.

Look at the picture of the lizards again. What questions do you have about this picture?

I see	
I think	
I wonder	

3.2 Light in Action | 65

3.2 | Wonder — How does light behave when it hits an object?

Activity 4
Observe Like a Scientist

Quick Code: us1536s

Shadows and Sunlight

Watch the video. **Look** for what happens when the animals are in sunlight. Then, **talk** about what you observe.

Video

Shadows and Sunlight

Talk Together

Now, talk together about shadows. Talk about what causes a shadow.

Activity 5
Observe Like a Scientist

Quick Code: us1537s

Shadow Play

Watch the video. **Look** for how the friends play with shadows. Then, **talk** about the question.

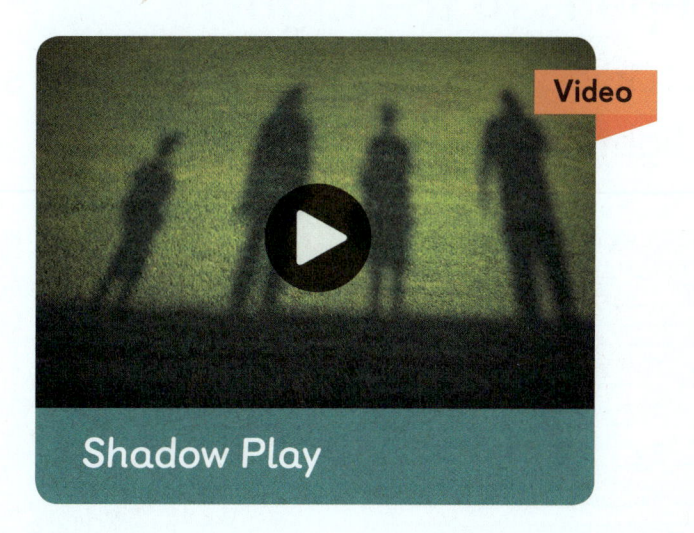

Shadow Play

Talk Together

Now, talk together about shadows. How can you change the size of shadows?

3.2 Light in Action | 67

3.2 | Wonder — How does light behave when it hits an object?

Activity 6
Evaluate Like a Scientist

Quick Code: us1538s

What Do You Already Know About Light in Action?

What Happens to the Light?

Look at each image. **Circle** the images that show an object that light will shine through.

How Does Light Travel?

Which photo shows light being transmitted?
Which shows light being reflected?
Draw a line to match each word to the correct photo.

Reflect

Transmit

3.2 Light in Action

3.2 | Learn How does light behave when it hits an object?

What Happens When Light Touches Something?

Activity 7
Observe Like a Scientist

Quick Code: us1539s

Windows with Shutters

Look at the picture. **Think** about the light inside and outside the house.

Windows with Shutters

Talk Together

Now, talk together. Would there be more light inside the house or outside the house?

Activity 8

Investigate Like a Scientist

Quick Code: us1540s

Hands-On Investigation: Making Shadow Puppets

In this activity, you will make puppets. Then you will shine **light** on your puppets to see what happens.

Make a Prediction

What will happen when you shine the light on the puppets?

Write or **draw** your predictions.

SEP Planning and Carrying Out Investigations
CCC Cause and Effect

3.2 Light in Action

3.2 | Learn How does light behave when it hits an object?

What materials do you need?
(per group)

- Small bedsheet
- Craft sticks
- Desk or table
- Flashlight
- Batteries, size D
- Glue sticks
- Construction paper
- Pencils
- Metric ruler
- Scissors
- Darkened area

What Will You Do?

Make your puppet using materials from your teacher. **Draw** your puppet.

Use the flashlight to shine light on your puppet from above, below, the left, and the right. **Write** or **draw** what happened.

3.2 | Learn — How does light behave when it hits an object?

Think About the Activity

Record which objects transmit light and which reflect light.

Object	Transmit or Reflect?

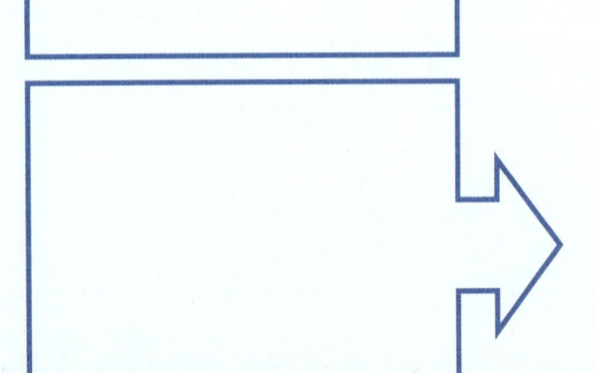

Which things let light pass through?

_____

Which things did not let light pass through?

_____

What happened when you shined the light on different materials?

Activity 9
Analyze Like a Scientist

Quick Code: us1541s

What Happens When Light Hits an Object?

Look at the picture. **Underline** the sentences that describe what happens when light hits this object.

Read Together

What Happens When Light Hits an Object?

Some objects do not allow light to pass through. They are **opaque**. Wooden shutters are opaque.

Light passes through windows. They are **transparent**.

When light hits a **translucent** object, some of the light shows through the object. The rest of the light is reflected or absorbed by the object.

3.2 Light in Action | 77

3.2 | Learn How does light behave when it hits an object?

Activity 10
Investigate Like a Scientist

Quick Code: us1542s

Hands-On Investigation: Letting Light Through

In this activity, you will shine light on different objects to see how much light comes through.

Make a Prediction

Which materials will light pass through?
Which materials will absorb the light?
Write or **draw** your predictions.

SEP Planning and Carrying Out Investigations
CCC Cause and Effect

What materials do you need? (per group)

- Flashlight
- Batteries, size D
- Flat mirror
- White copy paper, 1
- Construction paper
- Clear plastic lid
- Frosted plastic lid
- Table or desk
- Darkened area

How can you test to see how much light passes through objects? **Write** or **draw** your plan.

3.2 | Learn — How does light behave when it hits an object?

What Will You Do?

Shine the flashlight on each material. **Write** or **draw** what happened.

White Paper

Dark Paper

Clear Lid

Frosted Lid

Think About the Activity

Which things let the most light through?
Which things let some light through?
Which things let no light through?

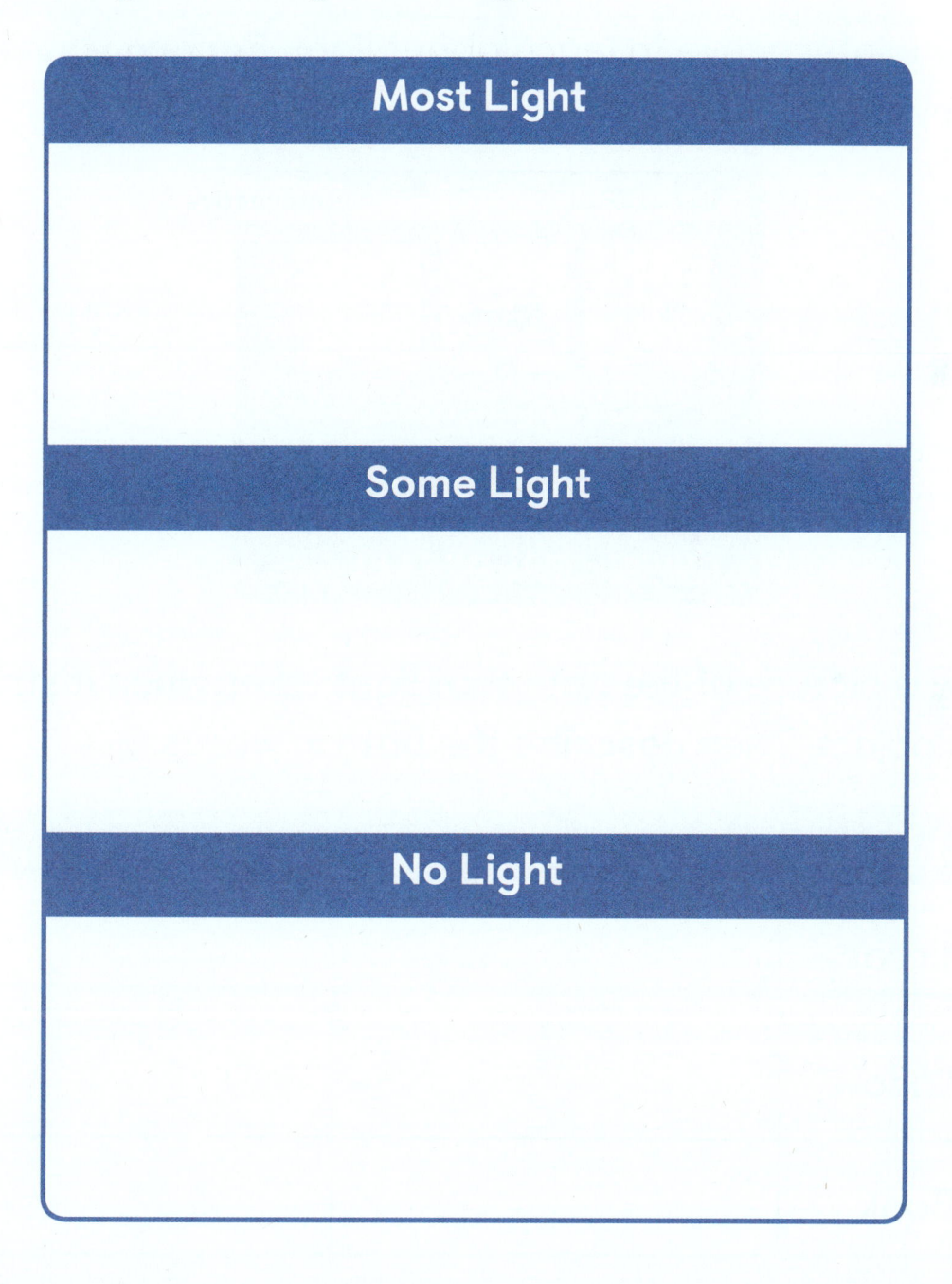

Most Light

Some Light

No Light

3.2 | Learn How does light behave when it hits an object?

Activity 11
Observe Like a Scientist

Quick Code: us1543s

Sound, Heat, and Light

Use the interactive to learn about light. **Record** your observations.

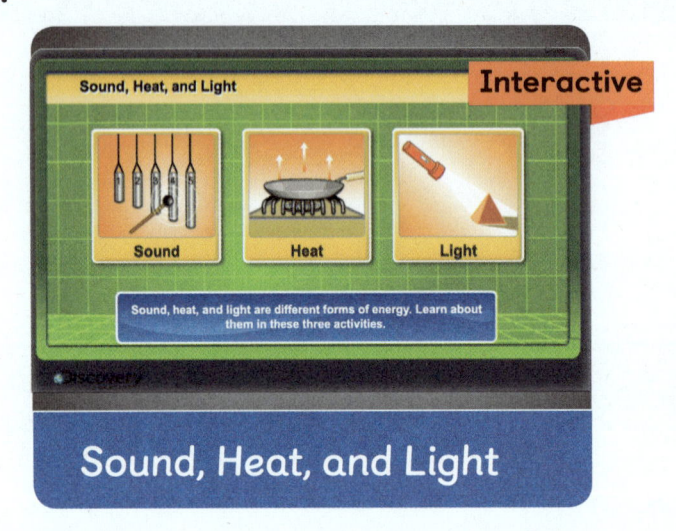

Sound, Heat, and Light

Draw a picture of the path that light takes when it strikes each object. Then **describe** the path in words.

Object	Draw	Describe
Mirror		
Window		
Book		

SEP Constructing Explanations and Designing Solutions

82

What do you think will happen when light strikes these objects? **Write** or **draw** what happens.

Object	What happens to light striking this object?
Basketball	
Paper for wrapping gifts	
Water in a wading pool	
The rearview mirror in a car	
Lens on a camera	
Mirror in a department store	
The shiny side of a CD	

3.2 Light in Action

Activity 12
Analyze Like a Scientist

Quick Code: us1544s

What Do You See?

Read about how light travels through objects. Then, **complete** the activity.

> Read Together

What Do You See?

Look out the window. You can see clearly through it. The window is transparent. Light easily travels through things that are transparent.

Transparent Window

SEP Planning and Carrying Out Investigations

Hold a thin piece of paper up to the light. You can see what is on the other side, but not very clearly. The paper is translucent. Some light can travel through things that are translucent.

Translucent Paper

Find a piece of metal. You can't see through it. The metal is opaque. Light can't travel through things that are opaque.

Opaque Metal

Find objects that are transparent, translucent, and opaque. **Draw** the objects in the chart.

Transparent	Translucent	Opaque

Activity 13

Observe Like a Scientist

Quick Code: us1545s

Looking at Light

Watch the video. Then, **talk** about the question.

Video

Light and Light Energy

Talk Together

Now, talk together. How did the shadow of the objects change when the position of the flashlight changed?

CCC Cause and Effect

3.2 Light in Action | 87

3.2 | Learn How does light behave when it hits an object?

Activity 14
Evaluate Like a Scientist

Quick Code: us1546s

Translucent, Transparent, or Opaque

Look at each object. **Draw** a line to show if each object is translucent, transparent, or opaque.

- Translucent 
- Transparent 
- Opaque

SEP Engaging in Argument from Evidence

How Can a Beam of Light Change Direction?

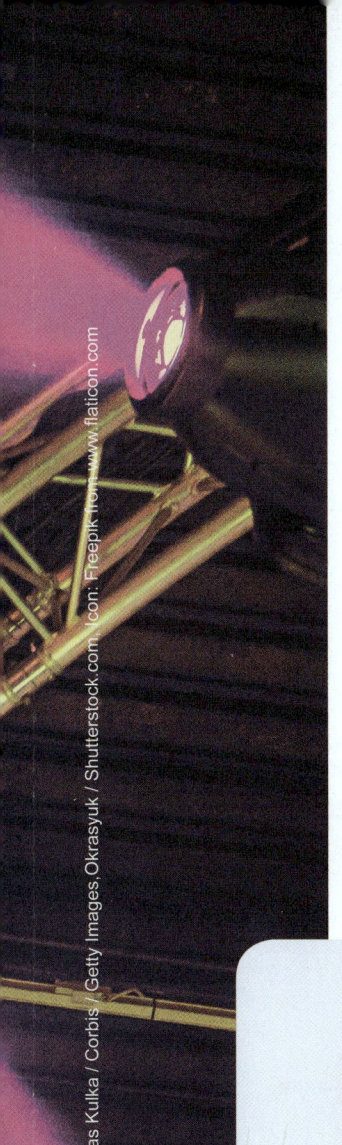

Activity 15
Analyze Like a Scientist

Quick Code: us1547s

How Can an Object Change the Direction of a Beam of Light?

Read about how light can change direction. **Think** about objects that reflect light.

Read Together

How Can an Object Change the Direction of a Beam of Light?

Can light be reflected onto other objects with a mirror? Yes it can. The light will bounce off the mirror and reflect to other objects.

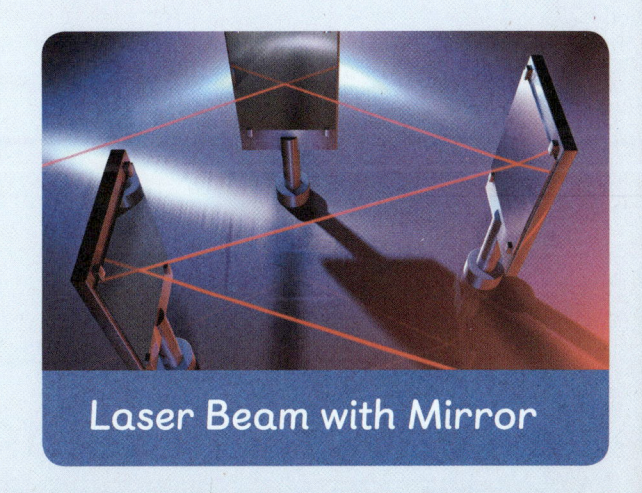

Laser Beam with Mirror

3.2 Light in Action | 89

Read Together

Looking in a Mirror

When light hits a mirror, it reflects back to your eyes. Objects that are flat and smooth reflect light very well.

Find objects that will reflect light. **Draw** the objects you find.

SEP Planning and Carrying Out Investigations

Activity 16
Observe Like a Scientist

Quick Code: us1548s

Scratched Mirror

Watch the video to learn about mirrors. Then, **talk** about the question.

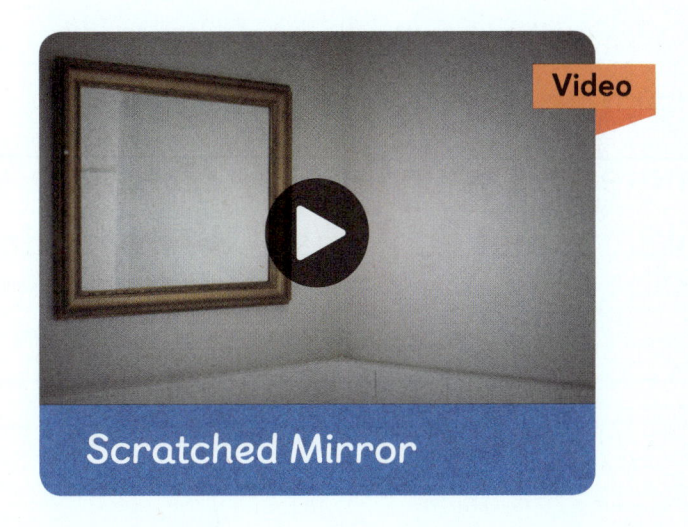

Scratched Mirror

Talk Together

Now, talk together about mirrors. How do mirrors work?

3.2 Light in Action | 91

3.2 | Learn How does light behave when it hits an object?

Activity 17
Investigate Like a Scientist

Quick Code: us1549s

Hands-On Investigation: Reflection

In this activity, you will see which materials are best at reflecting light.

Make a Prediction

Look at the materials your teacher gives you. Which do you think will reflect light best? **Write** or **draw** your answer.

What materials do you need?
(per group)

- Flashlight
- Batteries, size D
- Piece of marble
- Piece of painted metal
- Piece of shiny metal
- Flat mirror
- Copy paper
- Fake fur
- Plastic block
- Wooden blocks

SEP Planning and Carrying Out Investigations
CCC Cause and Effect

3.2 | Learn How does light behave when it hits an object?

What Will You Do?

How can you tell which materials will reflect the light best? **Write** or **draw** how you are going to test each material.

Use your plan to test three materials. **Write** or **draw** what you see.

Choose a material that reflects light. How can you use it to make a beam of light hit a target? **Write** or **draw** what you think.

Use your plan to make a beam of light hit the target. **Write** or **draw** what you see.

3.2 | Learn — How does light behave when it hits an object?

Think About the Activity

Material	What I Saw	Did It Reflect Light?

Which material was the best at reflecting light?
_____

How did you make your light hit the target?
_____
_____
_____

Activity 18
Observe Like a Scientist

Quick Code: us1550s

Reflection of Light

Watch the video to learn about how surfaces reflect light. Then, **complete** the sentence.

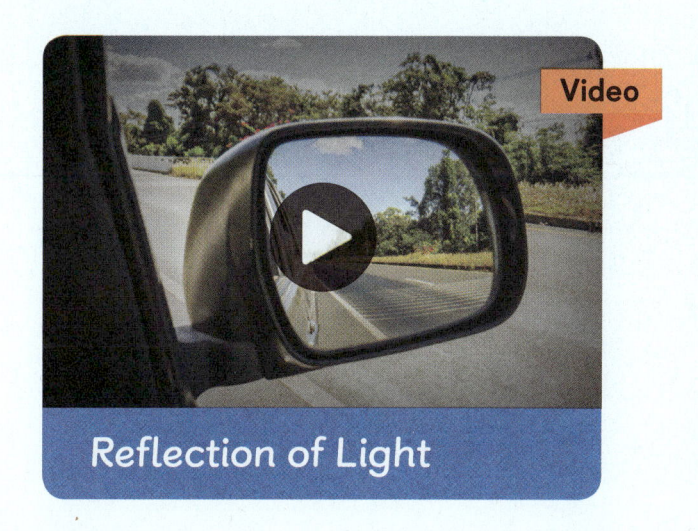

Reflection of Light

_____ surfaces scatter light toward our eyes while

_____ ones reflect light more _____ .

3.2 Light in Action

How does light behave when it hits an object?

Activity 19
Evaluate Like a Scientist

Quick Code: us1551s

Path of Light

at the picture.

and Effect

Read the words in the box. **Write** the words to **complete** each sentence about how light behaves in the picture.

| block | opaque | reflects | sun |

Light comes from the _____.

The mountains are _____.

The trees _____ some light.

The water _____ light.

3.2 | Share — How does light behave when it hits an object?

Activity 20
Record Evidence Like a Scientist

Quick Code: us1552s

Shadows

Now that you have learned about how light causes shadows, **look** again at the picture of the children playing. You first saw this in Wonder.

Let's Investigate Shadows

Talk Together

How can you describe shadows now? How is your explanation different from before?

SEP Constructing Explanations and Designing Solutions

Look at the Can You Explain? question. You first read this question at the beginning of the lesson.

> 💬 **Can You Explain?**
> How does light behave when it hits an object?

Now, you will use your new ideas about shadows to answer a question.

1. **Choose** a question. You can use the Can you Explain? question, or one of your own. You wrote questions at the beginning of the lesson.

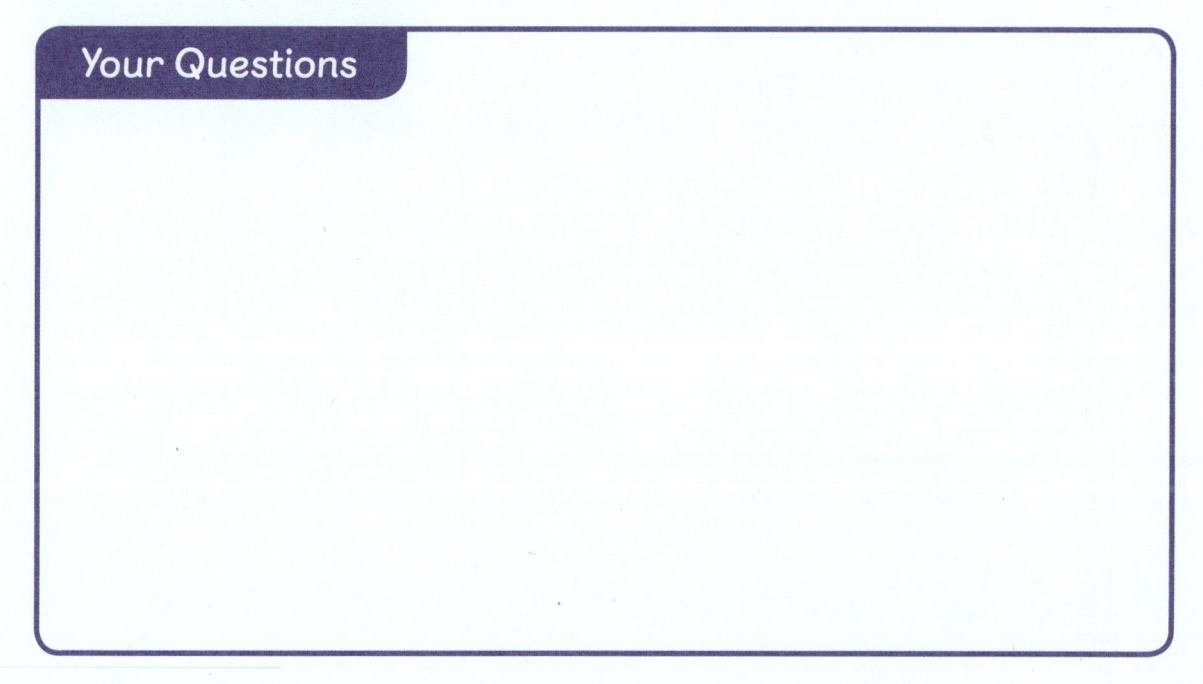

   Your Questions

2. **Use** the sentence starters to answer the question.

3.2 Light in Action

3.2 | Share How does light behave when it hits an object?

Light travels

_____
_____
_____

Materials that do not let light pass through them are

_____
_____

When light hits an opaque object

_____
_____
_____

Light changes direction when

_____
_____
_____

STEM in Action

Quick Code: us1553s

Activity 21
Analyze Like a Scientist

Transparent Aluminum

Circle the reasons glass is a good material to use.

Underline the reasons glass is not a good material to use.

Read Together

Transparent Aluminum

When you want to see through something, like a window or the screen of a cell phone, you need a transparent material. Most of the time the material is glass. Glass is clear, hard, and cheap. It is easy to make glass. But glass is brittle, which means it is easy to scratch and break. It is also very heavy.

Aluminum Cans

3.2 Light in Action

Read Together

Some plastics are transparent. Plastic is lighter and more flexible than glass. But plastic is not as strong as glass. Some metals are lighter and stronger than glass, but they are opaque.

Spinel is a new material. It is a clear ceramic made from aluminum. It is lighter and more flexible than glass. It is also more expensive and harder to make.

Dr. Jas Sanghera is an engineer who helped develop spinel. Dr. Sanghera started working for the U.S. Navy in 1988. He develops new materials that help people see and communicate better. These materials include new types of glass, crystal, ceramics, and fiber optics. On December 16, 2016, Dr. Sanghera won an award for his work developing new materials.

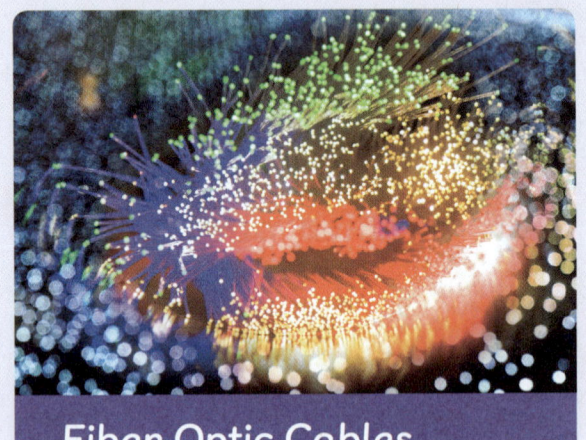

Fiber Optic Cables

The Right Stuff

Decide if each word describes a property of glass or spinel. Some words can describe both.

Write the words in the chart.

> strong   brittle   hard   cheap
> expensive   transparent   light   heavy

Glass	Spinel

SEP Obtaining, Evaluating, and Communicating Information

3.2 | Share — How does light behave when it hits an object?

Activity 22
Evaluate Like a Scientist

Quick Code: us1554s

Review: Light in Action

Think about what you have read and seen. What did you learn?

Draw what you have learned. Then, **tell** someone else about what you learned.

Talk Together

Think about what you saw in Get Started. Use your new ideas to discuss how different materials affect the path of light.

3.2 Light in Action | 107

CONCEPT
3.3

Sunlight

Student Objectives

By the end of this lesson:

☐ I can predict how the amount of sunlight will change during the year.

☐ I can find patterns in the amount of sunlight during different times of the day and year.

☐ I can find patterns in the amount of sunlight in different places.

Key Vocabulary

☐ axis
☐ Earth
☐ position
☐ rotate

Quick Code: us1556s

3.3 Sunlight

Activity 1
Can You Explain?

Does everyone on Earth have the same amount of sunlight?

_____

_____

_____

_____

_____

Quick Code: us1557s

3.3 Sunlight

3.3 | Wonder — Does everyone on Earth have the same amount of sunlight?

Activity 2
Ask Questions Like a Scientist

Quick Code: us1558s

Sun and Moon

Look at the photo. **Answer** the questions.

Let's Investigate Sun and Moon

SEP Asking Questions and Defining Problems

112

There is a story behind every picture.
What do you notice?

_____

Does the picture show day or night?

_____

What questions do you have about sunlight?

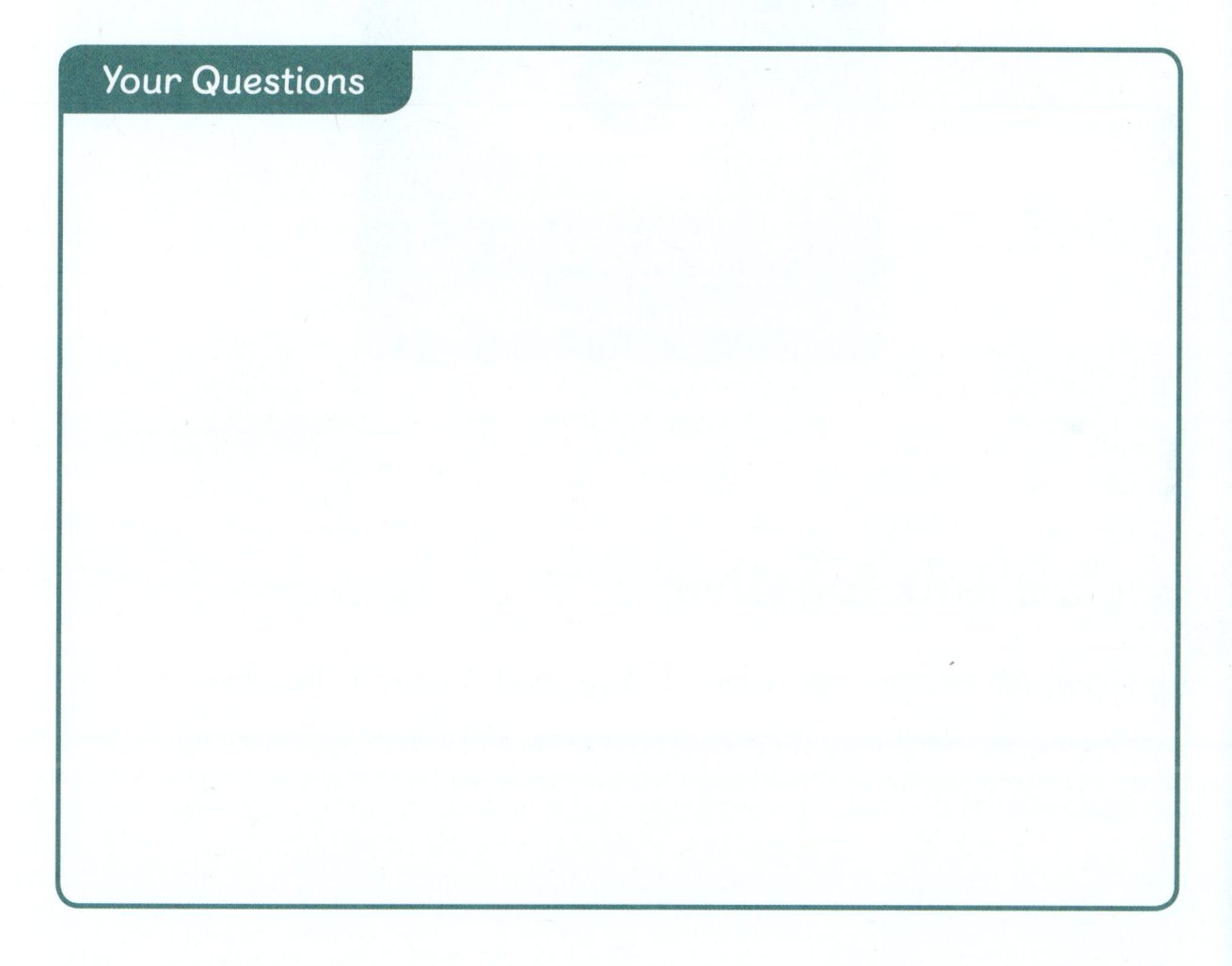

Your Questions

3.3 Sunlight | 113

3.3 | Wonder
Does everyone on Earth have the same amount of sunlight?

Activity 3
Observe Like a Scientist

Quick Code: us1559s

Describing Day and Night

Watch the video. Then, **talk** about the question.

Describing Day and Night (Video)

Talk Together

Now, talk together about day and night. Talk about how day and night are different. What causes day and night?

Activity 4

Evaluate Like a Scientist

Quick Code: us1560s

What Do You Already Know About Sunlight?

What Do You Know?

Think about the different things you can see in the sky.

Write or **draw** two things you see during the day.

Write or **draw** two things you see during the night.

Day	Night

3.3 Sunlight

3.3 | Wonder — Does everyone on Earth have the same amount of sunlight?

Picturing the Sky Above

Stargazing

Read each sentence.
Circle the sentences that are true.

- Stars do not exist during the day.

- The moon can only be seen at night.

- The sun seems to travel across the night sky.

- There are millions of stars that can be seen at night.

- The moon is a star.

- The sun is a star.

3.3 | Learn Does everyone on Earth have the same amount of sunlight?

What Are the Patterns of Sunlight?

Activity 5
Observe Like a Scientist

Quick Code: us1561s

Shell Beach

Look at the picture. Then, **talk** about the question.

Shell Beach

Talk Together

Now talk together. What time of day do you think it is? Can you tell the time of day by looking at the shadows?

Activity 6

Investigate Like a Scientist

Quick Code: us1562s

Hands-On Investigation: Measuring Shadows

In this activity, you will see how shadows change during the day.

Make a Prediction

Think about how the sun moves during the day. **Think** about how shadows change when the sun is high and low in the sky.

Write or **draw** how shadows will change during the day.

| SEP | Using Mathematics and Computational Thinking |
| CCC | Patterns |

3.3 Sunlight | 119

3.3 | Learn   Does everyone on Earth have the same amount of sunlight?

What materials do you need?
(per group)

- Large sheets of paper
- Markers
- Chalk
- Scissors
- Clock

What Will You Do?

Go outside on a sunny day. **Trace** your partner's shadow on the large paper. **Write** the date and time. **Draw** your partner, the shadow, and the sun in the box below.

Wait a while. **Trace** your partner's shadow on the large paper again. **Write** the date and time. **Draw** your partner, the shadow, and the sun in the box below.

Wait a while. **Trace** your partner's shadow on the large paper one more time. **Write** the date and time. **Draw** your partner, the shadow, and the sun in the box below.

3.3 | Learn — Does everyone on Earth have the same amount of sunlight?

Think About the Activity

Look at your drawings. **Write** or **draw** how your shadow changed.

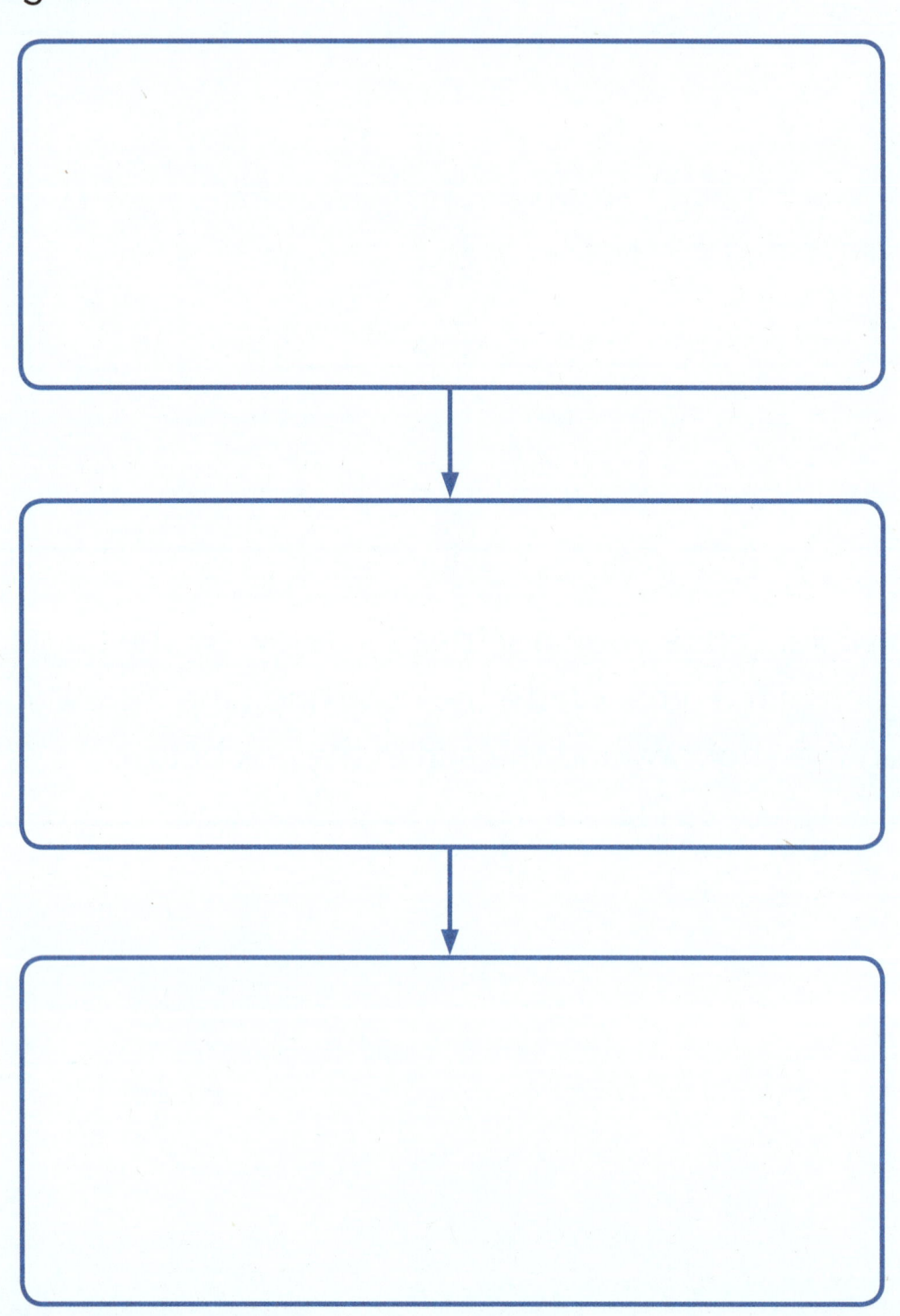

What patterns did you see?

___
___
___

How do you think the changes in the shadow relate to the sun's **position** in the sky?

___
___
___

What do you think the changes in shadows mean about how **Earth** moves?

___
___
___

3.3 | Learn
Does everyone on Earth have the same amount of sunlight?

Activity 7
Observe Like a Scientist

Quick Code: us1563s

Sun across the Sky

Look at the picture.

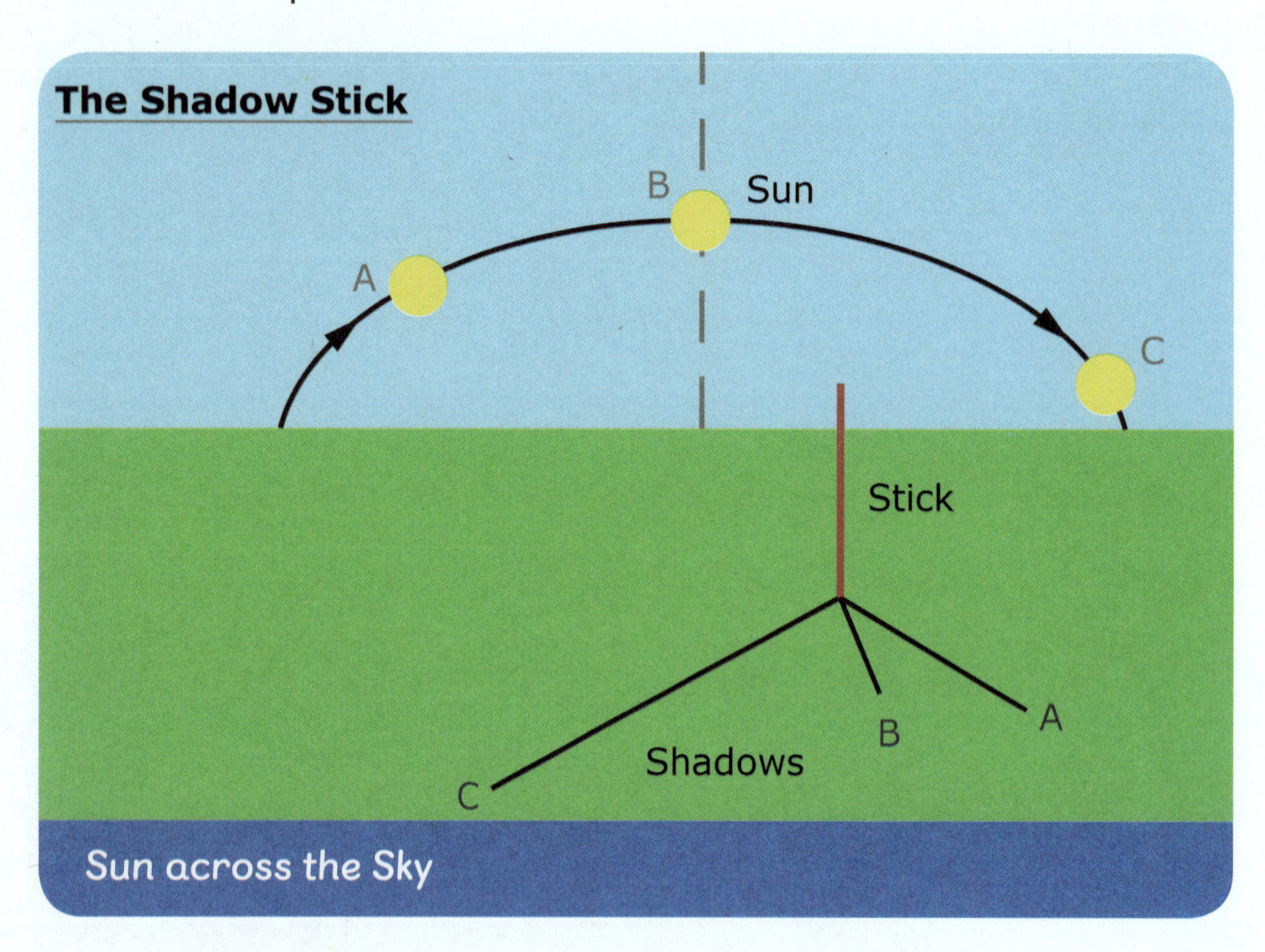

The Shadow Stick

Sun across the Sky

SEP Developing and Using Models

Look at the pattern of the shadows made by the stick.

How is this pattern similar to the shadows you traced? **Write** or **draw** your answer.

How is this pattern different from the shadows you traced? **Write** or **draw** your answer.

3.3 Sunlight | 125

3.3 | Learn Does everyone on Earth have the same amount of sunlight?

Activity 8
Observe Like a Scientist

Quick Code: us1564s

Exploring Shadows

Watch the video. **Talk** together about what you saw in the video. Then, **answer** the questions.

Exploring Shadows

Talk Together

Now, talk together about how shadows change when the angle changes.

SEP Analyzing and Interpreting Data

How does the bottle's shadow change when the angle is lowered? **Write** or **draw** your answer.

How is this pattern similar to the shadows you traced? **Write** or **draw** your answer.

How is this pattern different from the shadows you traced? **Write** or **draw** your answer.

3.3 Sunlight | 127

Read Together

Shadows

Your shadow is made by sunlight on your body. The sunlight does not go through you. That is why it forms a shadow. The shadow appears on the other side of where light comes from.

Shadows on a Wall

The sun makes your shadow. The sun appears to move across the sky through the day. That is what makes your shadow different lengths through the day.

Your shadow is the longest after sunrise and just before sunset. In the middle of the day your shadow is very short. It is so short you could step on your whole shadow.

Activity 9

Analyze Like a Scientist

Quick Code: us1565s

Shadows

Look at the times from your investigation. At what time was the shadow longest? At what time was the shadow shortest?

Do these times match the times in the passage?

_____

What would your shadow look like if you measured it at sunrise? **Write** or **draw** your answer.

| SEP | Engaging in Argument from Evidence |

3.3 Sunlight | 129

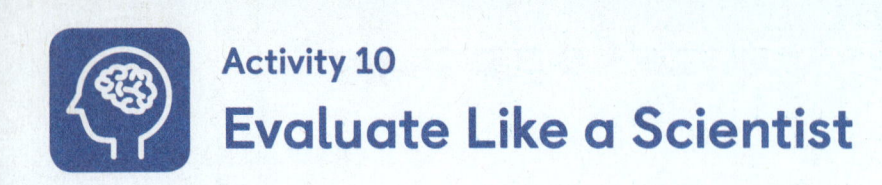

3.3 | Learn Does everyone on Earth have the same amount of sunlight?

Activity 10
Evaluate Like a Scientist

Quick Code: us1566s

Graham's Shadow Experiment

Graham looked at sun and shadows. He placed a toy building in the sun. During the day he measured the length of its shadow.

Look at the graphs. **Circle** the graph that shows the length of the toy's shadow during the day.

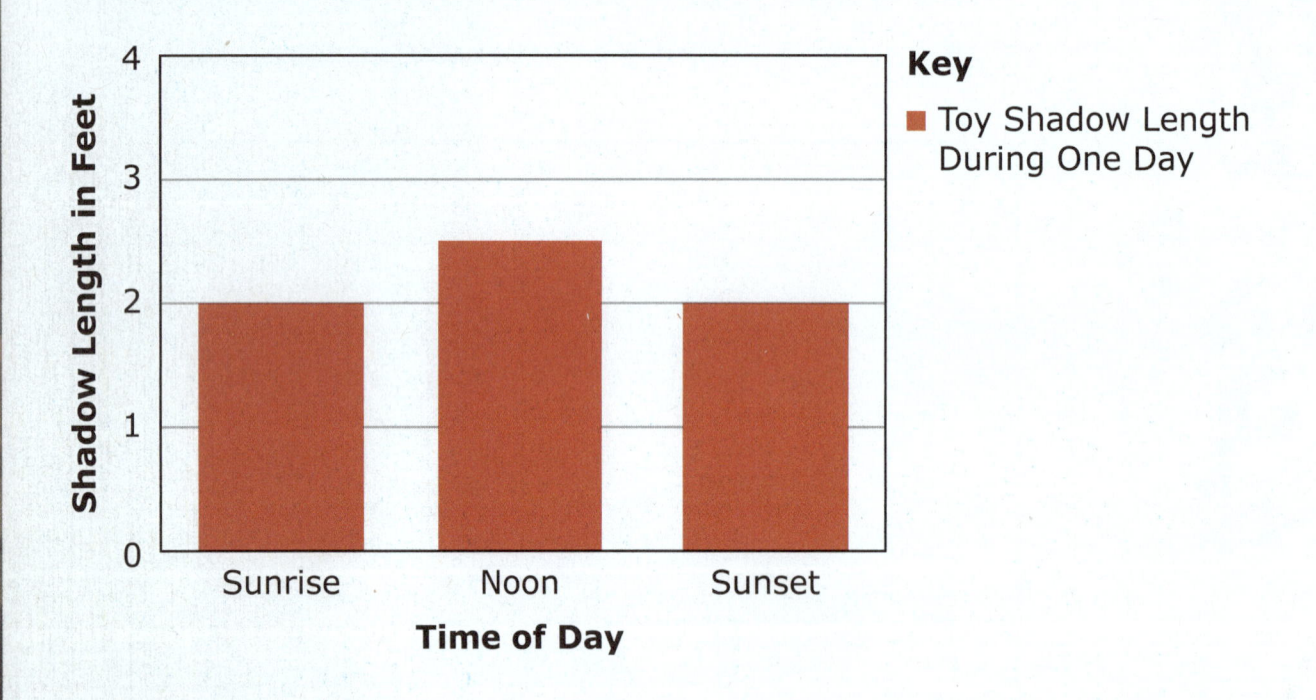

Key
- Toy Shadow Length During One Day

Shadow Length in Feet

Sunrise — 2
Noon — 2.5
Sunset — 2

Time of Day

SEP Analyzing and Interpreting Data

130

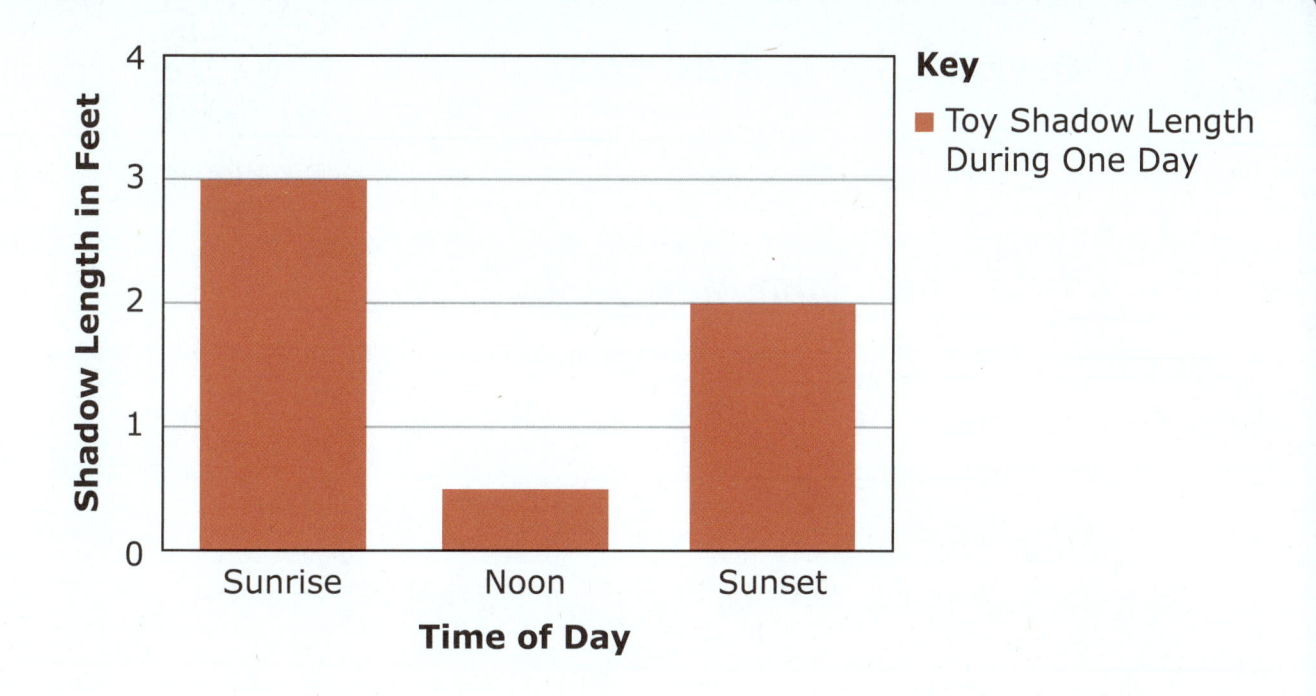

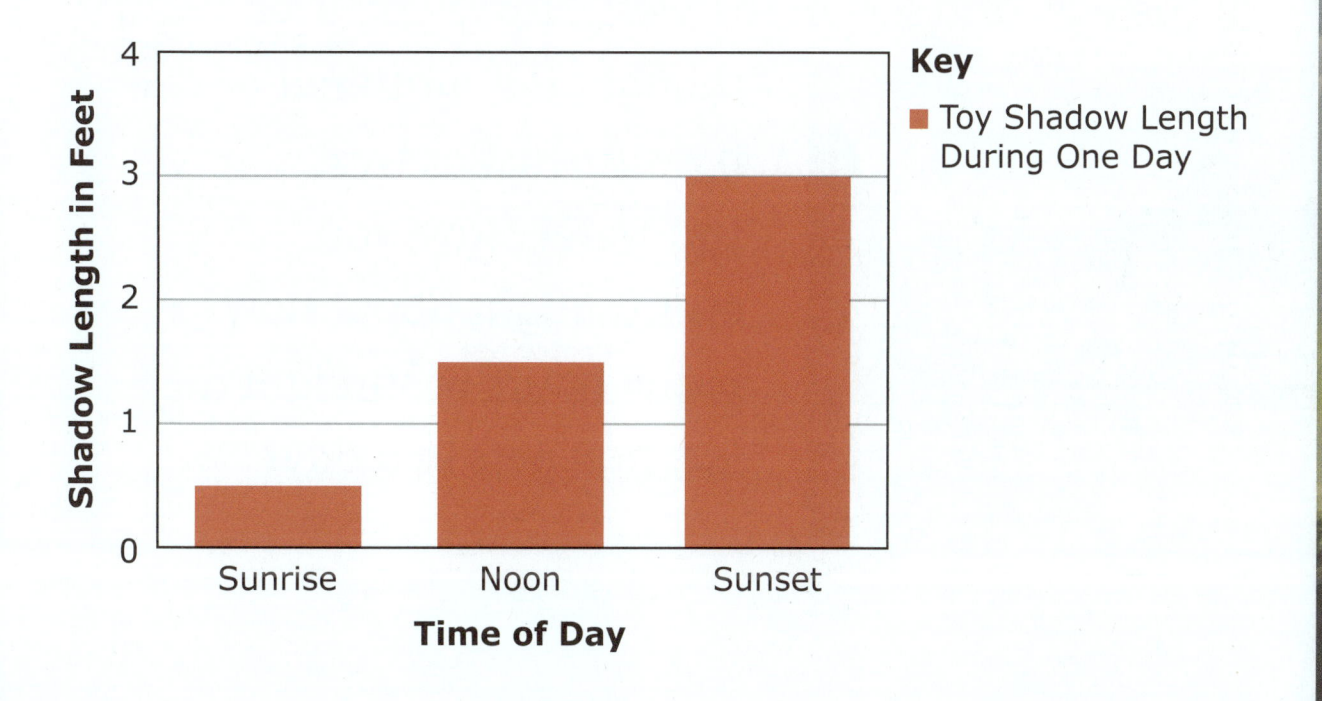

Shadow Length in Feet (Chart 1)

Time of Day	Shadow Length (ft)
Sunrise	3
Noon	0.5
Sunset	2

Key: Toy Shadow Length During One Day

Shadow Length in Feet (Chart 2)

Time of Day	Shadow Length (ft)
Sunrise	0.5
Noon	1.5
Sunset	3

Key: Toy Shadow Length During One Day

3.3 Sunlight

Activity 11
Analyze Like a Scientist

Quick Code: us1567s

Sunrise and Sunset

Read the article. Then, **complete** the activity.

> Read Together

Sunrise and Sunset

You've probably heard the saying "The sun rises in the east and sets in the west." Have you ever thought if it mattered how far east or west you were? Have you ever thought about how many hours of sunlight a location gets in a day?

Sunrise

SEP Analyzing and Interpreting Data

The time that the sun rises each day is called sunrise. What does it mean for the sun to rise? Think of how dark it is at night. Then, think of the first moment that the sun appears in the early morning. The sun "rises" into your view on the horizon. If you watch the sun rise, you will notice it happens to the east. And it happens every day.

The time that the sun sets each day is called sunset. What does it mean for the sun to set? Think of the end of the day when you can barely see the sun. It is the moment when you cannot see the sun anymore on the horizon. It is like it has disappeared. Where does the sun go when you cannot see it? If you watch the sun set, you will notice that it happens to the west. And it happens every day.

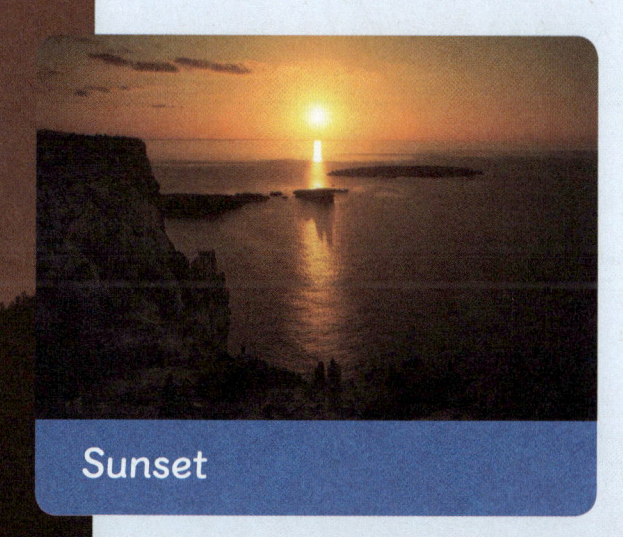

You can find the number of hours of sunlight by finding the difference between sunset and sunrise times.

Sunset

Look at the sunrise and sunset times for London, Seattle, and Ocho Rios on July 21 and December 21.

Date	Location	Sunrise (local time)	Sunset (local time)
July 21	Seattle	5:34 AM	8:57 PM
July 21	London	5:09 AM	9:04 PM
July 21	Ocho Rios	5:43 AM	6:47 PM
December 21	Seattle	7:55 AM	4:20 PM
December 21	London	8:04 AM	3:54 PM
December 21	Ocho Rios	6:36 AM	5:38 PM

Circle the location that gets the most sunlight in July. **Underline** the location that gets the most sunlight in December.

Look at the graph. The bars show hours of sunlight in each city for the two different dates. Use the information in the table and the graph to **answer** the questions.

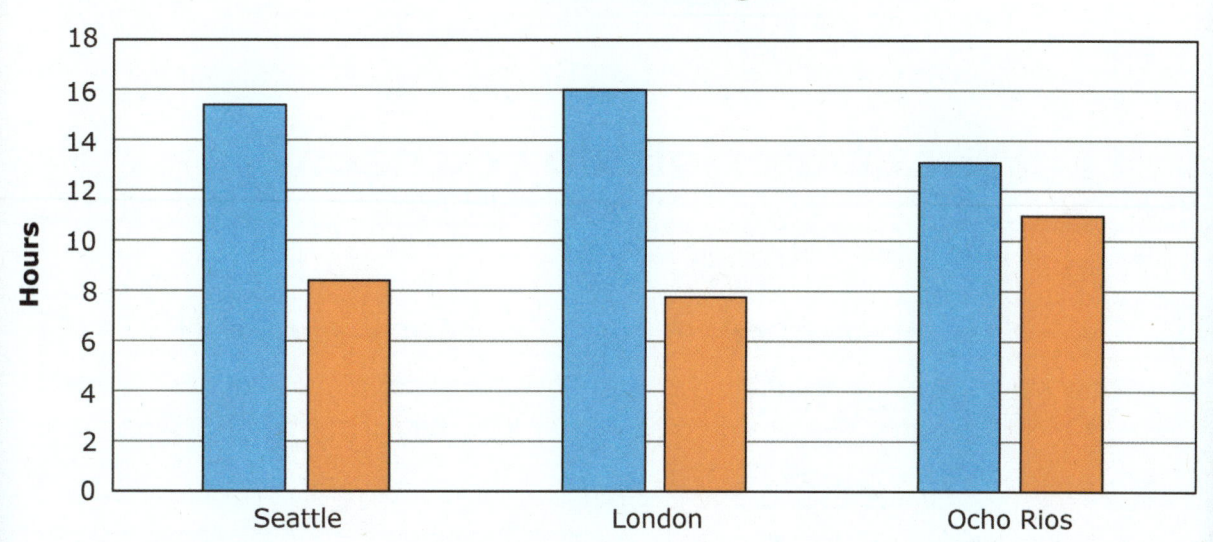

Hours of Sunlight

City	Blue	Orange
Seattle	~15.5	~8.5
London	16	~7.5
Ocho Rios	13	11

What date is shown by the blue bars?

_____

What date is shown by the orange bars?

_____

3.3 Sunlight | 135

3.3 | Learn Does everyone on Earth have the same amount of sunlight?

Activity 12
Think Like a Scientist

Quick Code: us1568s

Daylight Where I Live

In this activity, you will look for patterns in sunrise and sunset. You will collect data about the sun where you live.

What materials do you need?
(per group)

- Sunrise and sunset data
- Large three-column chart
- Paper plate
- Metal brads, 6
- Construction paper
- Markers

SEP Analyzing and Interpreting Data
CCC Patterns

What Will You Do?

Look at the sunrise and sunset data your teacher gives you. Do you see any patterns?

_____
_____
_____

Plan how to collect sunrise and sunset data for your area.

_____
_____
_____

Write the sunrise and sunset data for your area at least a week. **Repeat** for two months. How can you show this data?

_____
_____

3.3 | Learn Does everyone on Earth have the same amount of sunlight?

Date	Sunrise Time	Sunset Time

Date	Sunrise Time	Sunset Time

3.3 | Learn — Does everyone on Earth have the same amount of sunlight?

Date	Sunrise Time	Sunset Time

Think About the Activity

Find a pattern in the data from your teacher.

Find a pattern in the data you collected.

Data from My Teacher	My Data

How are the patterns similar? How are they different?

_____

_____

_____

3.3 Sunlight | 141

Activity 13

Analyze Like a Scientist

Quick Code: us1569s

Moving Sun

Read about the sun. **Underline** the sentence that explains why the sun appears to move.

Read Together

Moving Sun

Have you ever noticed that the sun rises on one side of the sky and sets on another side of the sky? Do you think the sun is moving?

Sunrise

The sun does not move. Earth **rotates**, or spins, as it travels around the sun.

Earth's rotation causes day and night. The side of Earth that faces the sun has day. The side of Earth that faces away from the sun has night.

Activity 14
Observe Like a Scientist

Quick Code: us1570s

Cycles in the Sky

Watch the Earth and Sun part of the interactive.

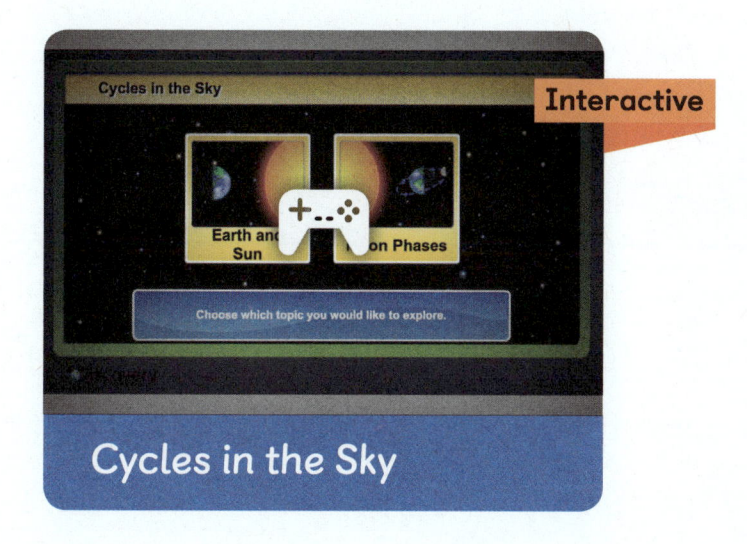

SEP Developing and Using Models

3.3 Sunlight | 143

3.3 | Learn Does everyone on Earth have the same amount of sunlight?

Draw a picture of the sun and Earth that shows day and night on Earth. **Shade** the part of Earth that shows night.

What is Earth's **axis**? **Label** Earth's axis in your drawing.

_____

_____

_____

Imagine it is noon where you live. Where on Earth would it be midnight? Why do you think so?

_____
_____
_____
_____

Do you think Earth rotates clockwise or counterclockwise? Why do you think so?

_____
_____
_____
_____

3.3 Sunlight | 145

3.3 | Learn — Does everyone on Earth have the same amount of sunlight?

Activity 15
Observe Like a Scientist

Quick Code: us1571s

Day and Night

Watch the video. **Look** for what causes day and night. **Talk** about the question.

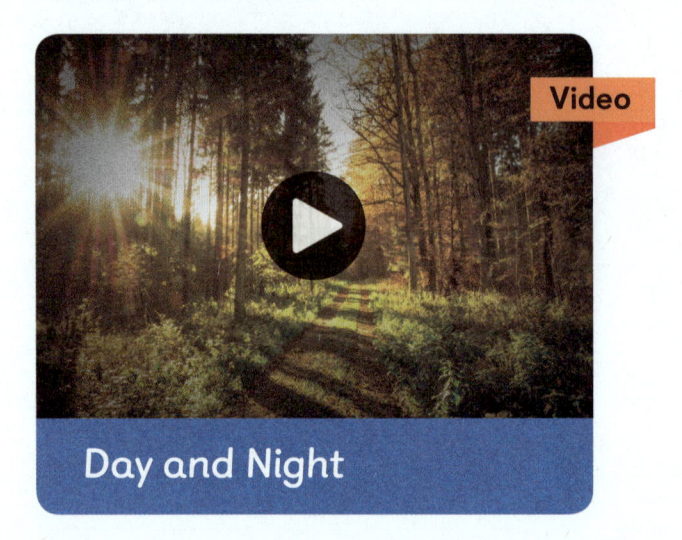

Video: Day and Night

Talk Together

Now, talk together about what causes day and night. How could you make a skit showing this?

SEP Obtaining, Evaluating, and Communicating Information

146

Work with a group to **plan** a skit about Earth's rotation, day, and night.

Use this chart to plan your skit.

Skit Planner	
What Tools Will We Need?	What Sights Will We See?
What Problem Might We Find?	How Can We Solve the Problem?

3.3 Sunlight

3.3 | Learn — Does everyone on Earth have the same amount of sunlight?

Activity 16
Evaluate Like a Scientist

Quick Code: us1572s

Day and Night on Earth

Earth is always light in one place and dark in another. What does Earth look like from above during day and night?

Sunset

Look at each picture. **Draw** a line from each picture to the part of Earth where it is that time of day.

3.3 Sunlight | 149

3.3 | Share Does everyone on Earth have the same amount of sunlight?

Activity 17
Record Evidence Like a Scientist

Quick Code: us1573s

Sun and Moon

Now that you have learned about patterns of sunlight, look again at the picture Sun and Moon. You first saw this in Wonder.

Let's Investigate Sun and Moon

Talk Together

How can you describe Sun and Moon now? How is your explanation different from before?

SEP Constructing Explanations and Designing Solutions

Look at the Can You Explain? question. You first read this question at the beginning of the lesson.

> ### Can You Explain?
> Does everyone on Earth have the same amount of sunlight?

Now, you will **use** your new ideas about Sun and Moon to answer a question.

1. **Choose** a question. You can use the Can you Explain? question, or one of your own. You wrote questions at the beginning of the lesson.

   Your Questions

2. Then, **use** the sentence starters on the next page to answer the question.

3.3 Sunlight | 151

3.3 | Share Does everyone on Earth have the same amount of sunlight?

I know the picture shows

_____

_____

because

_____

_____

This area will get fewer hours of sunlight because

_____

_____

An example of patterns in the amount of sunlight is

_____

_____

_____

Another example is

_____

_____

_____

The evidence I collected shows

_____

_____

_____

_____

STEM in Action

Quick Code: us1574s

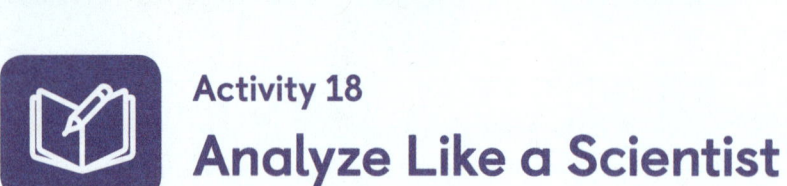

Activity 18
Analyze Like a Scientist

Discovering Day and Night

Read the passage. **Watch** the video. Then, **answer** the question.

Read Together

Discovering Day and Night

Earth turns on its axis, causing a full cycle of day and night every 24 hours. Airline pilots need to know how this happens. Flying planes all over the world means that pilots need to know when day starts and night ends. This way, they can be on time with their flights.

SEP Obtaining, Evaluating, and Communicating Information
SEP Using Mathematics and Computational Thinking

Because Earth spins on its axis, the time in one part of the country is not the same time as another. It may be daytime where you are, but it is also nighttime on another part of Earth. Airline pilots need to know this when they are flying planes from one place to another.

Hours and Minutes (Video)

How can you tell time with a clock?

_____
_____
_____
_____

Pilot for a Day

Imagine you are an airplane pilot. You are trying to figure out if you can make it home for dinner in Austin, Texas. Your plane leaves Los Angeles at 1:00 p.m. You are flying to Austin. The flight takes three hours. It is two hours later in Austin because of the time difference.

Circle the clock that shows the time you will land in Austin.

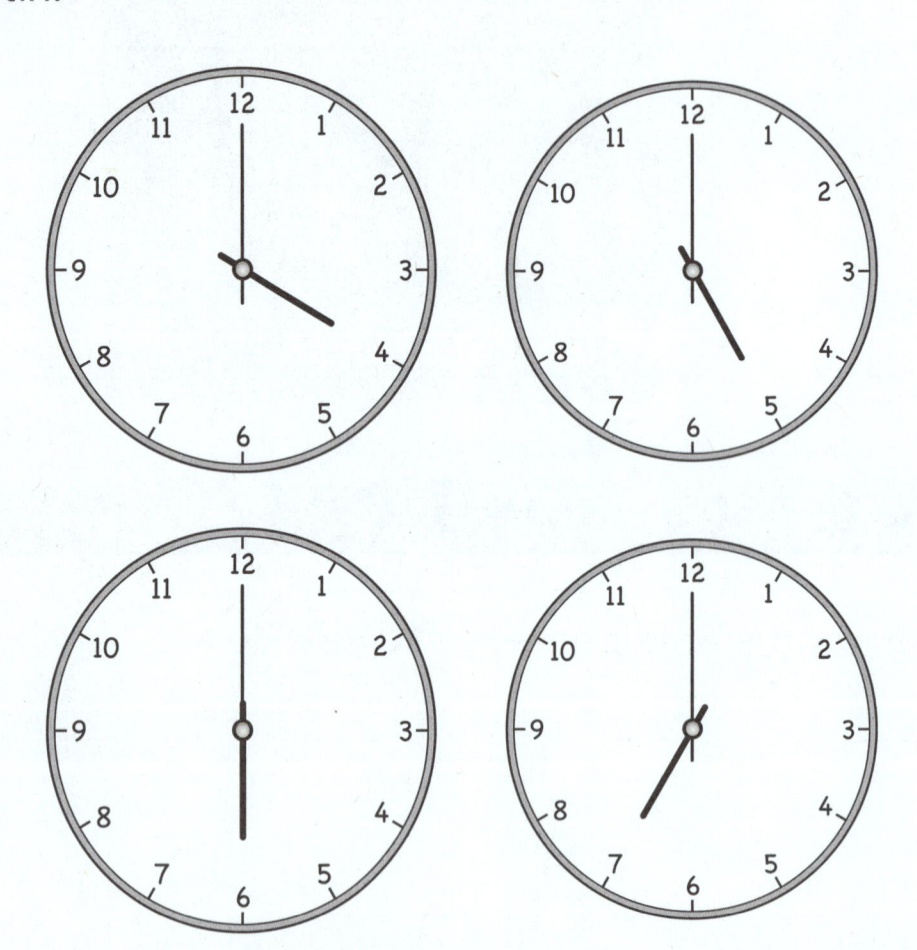

Activity 19

Evaluate Like a Scientist

Quick Code: us1575s

Review: Sunlight

Think about what you have read and seen. What did you learn?

Draw what you have learned.

Then, **tell** someone else about what you learned.

Talk Together

Think about what you saw in Get Started. Use your new ideas to discuss patterns in sunlight.

SEP Obtaining, Evaluating, and Communicating Information

3.3 Sunlight | 157

CONCEPT 3.4

Moon on the Move

Student Objectives

By the end of this lesson:

- ☐ I can make a model to show the patterns in the movement of the moon.
- ☐ I can find and describe patterns in the movement of stars at night.
- ☐ I can talk about patterns in the night sky.

Key Vocabulary

- ☐ binoculars
- ☐ constellation
- ☐ moon
- ☐ orbit

Quick Code: us1577s

3.4 Moon on the Move

Activity 1
Can You Explain?

How do objects in the night sky move?

_____
_____
_____
_____

Quick Code:
us1578s

3.4 Moon on the Move | 161

3.4 | Wonder
How do objects in the night sky move?

Activity 2
Ask Questions Like a Scientist

Quick Code: us1579s

The Night Sky

Look at the photo. **Answer** the questions.

Let's Investigate the Night Sky

162

There is a story behind every picture. What do you notice?

_____
_____
_____

What do you see in the sky? Does this look like night or day?

_____
_____
_____

What questions do you have about objects in the night sky?

Your Questions

3.4 Moon on the Move | 163

3.4 | Wonder — How do objects in the night sky move?

Activity 3
Observe Like a Scientist

Quick Code: us1580s

Astronomy

Complete the interactive. **Look** at objects in the night sky with a telescope and **binoculars**.

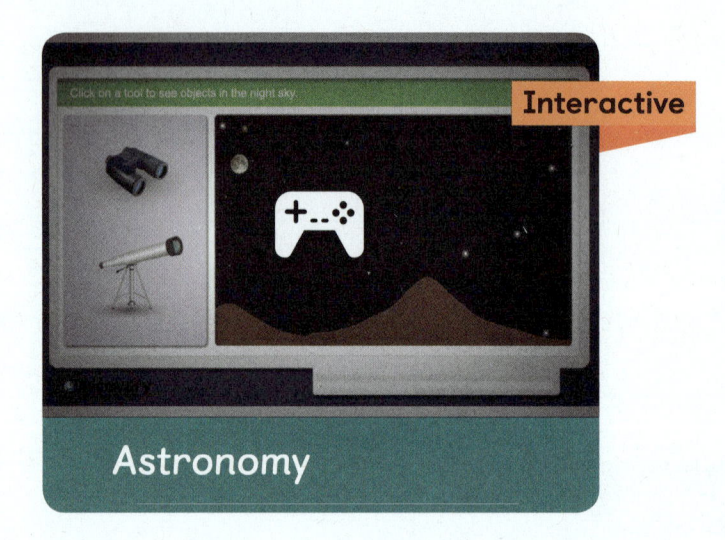

Interactive

Astronomy

Draw how the **moon** looked when you looked at it through the binoculars.

Draw how the moon looked when you looked at it through the telescope.

Binoculars	Telescope

3.4 Moon on the Move

3.4 | Wonder — How do objects in the night sky move?

Activity 4
Evaluate Like a Scientist

Quick Code: us1581s

What Do You Already Know About Moon on the Move?

What Is in the Sky Above?

Think about what you might see in the sky when you look up. **Read** each sentence. **Circle** the one that is correct.

- The clouds are in space, not in Earth's atmosphere.

- The stars are something you can only see during the day.

- The moon goes through phases, and you can see this happening each month when you look at the sky.

- The sun is closer to Earth than the moon.

Moon and Clouds

Sun, Moon, and Earth

Objects we see in the night sky appear larger or smaller based on how far away they are from Earth.

Pretend the objects are all the same distance from Earth. **Circle** the largest. **Draw** a square around the smallest.

Jupiter

Sun

Cloud

Moon

3.4 Moon on the Move | 167

3.4 | Learn How do objects in the night sky move?

What Patterns Do We See in the Sky at Night?

Activity 5
Observe Like a Scientist

Quick Code: us1582s

What We See at Night

Watch the video to observe objects we see at night.

Video

What We See at Night

Talk Together

Now, talk together about objects we see in the night sky.

168

Activity 6

Think Like a Scientist

Quick Code: us1583s

Star Maker

In this activity, you will make a star **constellation** that you can see any time you want. A constellation is a group of stars. They form a pattern in the night sky.

What materials do you need?
(per group)

- Construction paper
- Pencils
- Scissors
- Clear tape
- Tube, cardboard, toilet paper roll

What Will You Do?

Choose a constellation to make. **Read** to learn more about your constellation.

| SEP | Developing and Using Models |
| CCC | Patterns |

3.4 Moon on the Move | 169

3.4 | Learn How do objects in the night sky move?

Draw your constellation.

Write what you learned about your constellation.

_____

_____

_____

_____

1. **Cut** a circle out of black construction paper.

2. **Draw** dots on the black circle to show where the stars in the constellation should be.

3. **Cover** one end of the cardboard tube with the paper. The side with the dots should be facing out. **Tape** the paper to the tube.

4. Carefully **use** the pencil to poke holes where the dots are.

5. **Hold** your tube up to a light source. **Write** or **draw** what you see.

3.4 | Learn How do objects in the night sky move?

Think About the Activity

How does your model compare to an actual constellation?

_____

_____

_____

_____

Activity 7

Analyze Like a Scientist

Quick Code: us1584s

Sun and Stars

Look at the picture. Which sentence in the paragraph describes this picture? **Circle** the sentence.

Read Together

Sun and Stars

Stars are big balls of burning gas. They are scattered unevenly in space. Stars look small because they are far away.

The sun is also a star. It is closer to Earth than any other star. That's why it looks big and bright.

A Tent and the Night Sky

SEP Engaging in Argument from Evidence

3.4 Moon on the Move

Is the sun part of a constellation? Why or why not?

_____

_____

_____

Activity 8
Evaluate Like a Scientist

Quick Code: us1585s

Objects in the Sky

Look at the pictures of objects in the sky. **Circle** the pictures that show the sun. **Draw** a square around the pictures that show constellations.

SEP Analyzing and Interpreting Data

3.4 Moon on the Move | 175

Activity 9
Analyze Like a Scientist

Quick Code: us1586s

Moon During the Day

Circle the sentence that describes the picture.

> Read Together

Moon During the Day

The sun is the only star we see during the day. Other stars we can only see at night. We can see the moon during the day and night. If we can see the moon during the day depends on where the moon is in its **orbit**.

Moon During the Day

SEP Constructing Explanations and Designing Solutions

Activity 10
Observe Like a Scientist

Quick Code: us1587s

Stars Moving

Look at the picture. Then, **answer** the question.

Stars Moving

Why do the stars in this picture look like lines?

_____

_____

_____

SEP Constructing Explanations and Designing Solutions

3.4 Moon on the Move | 177

3.4 | Learn How do objects in the night sky move?

Activity 11
Observe Like a Scientist

Quick Code: us1588s

Cycles in the Sky

The moon looks like it moves across the sky at night.

Explore cycles of the moon online.

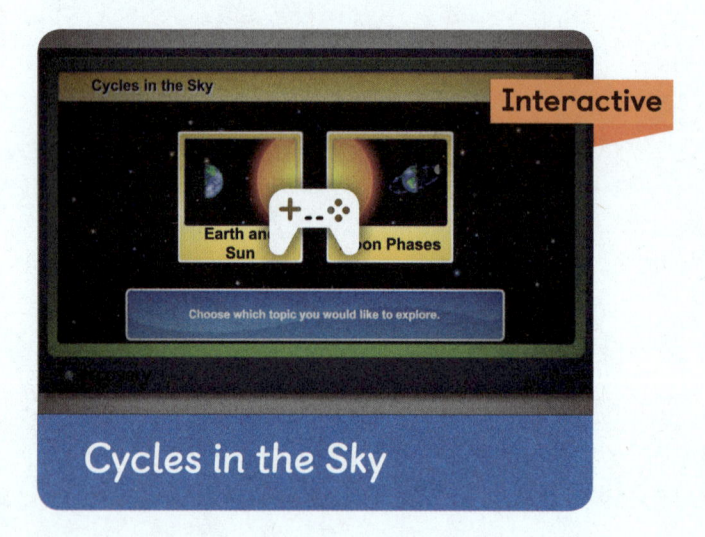

Cycles in the Sky
Interactive

SEP Developing and Using Models

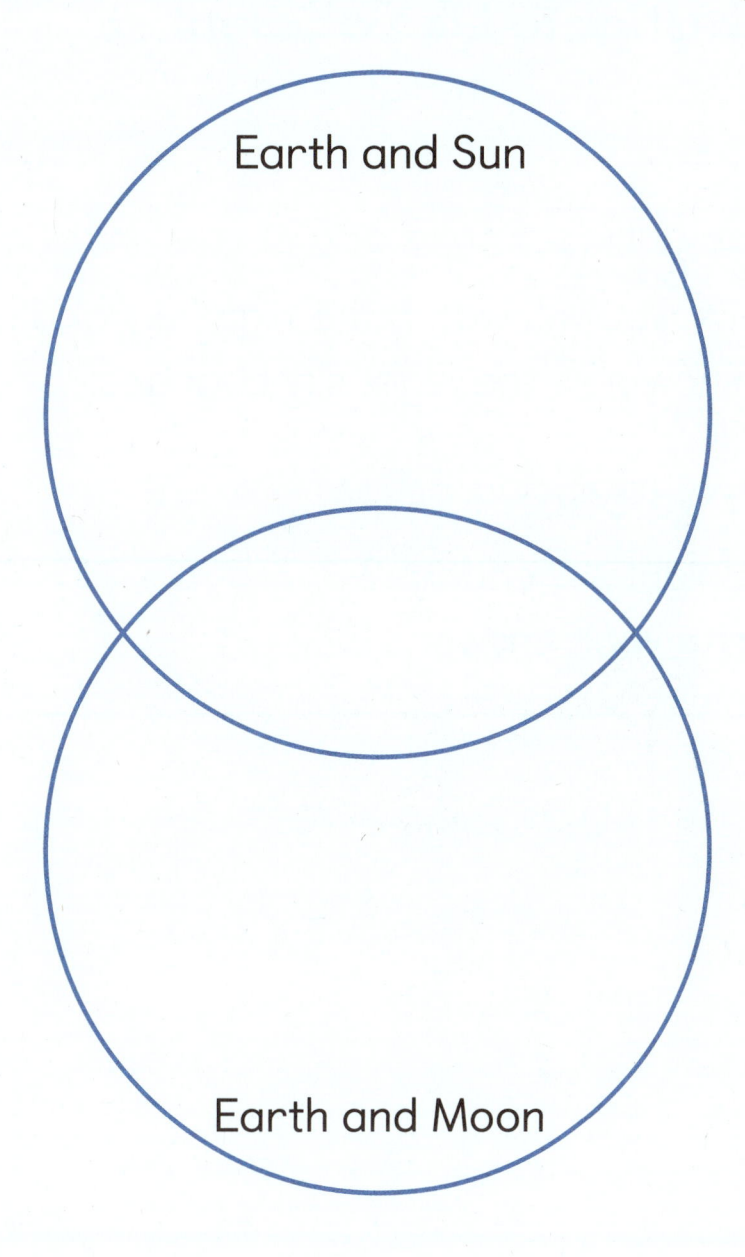

Compare patterns of the sun with patterns of the moon.

Earth and Sun

Earth and Moon

3.4 Moon on the Move

3.4 | Learn How do objects in the night sky move?

Activity 12
Investigate Like a Scientist

Quick Code: us1589s

Hands-On Investigation: Night Observations

In this activity, you will watch the night sky and draw pictures to show how the night sky changes.

Make a Prediction

How will the night sky change?
Draw or **write** your answer.

SEP Analyzing and Interpreting Data
CCC Patterns

What materials do you need?
(per group)

- Blanket or chair
- Flashlight
- Batteries, size D
- Sheets of paper, 4
- Colored pencils
- Clipboard
- Watch

HANDS-ON INVESTIGATION

What Will You Do?

Watch the sky at night. **Write** the date and time. What did you see? **Draw** what you see on a piece of paper.

3.4 Moon on the Move

3.4 | Learn How do objects in the night sky move?

Wait 15 minutes. **Write** the time. What did you see? Then, **look** at the sky again. **Draw** what you see.

Wait two weeks. **Write** the date and time. **Watch** the sky at night. **Draw** what you see. What did you see?

Wait 15 minutes. **Write** the date and time. Then, **look** at the sky again. **Draw** what you see. What did you see?

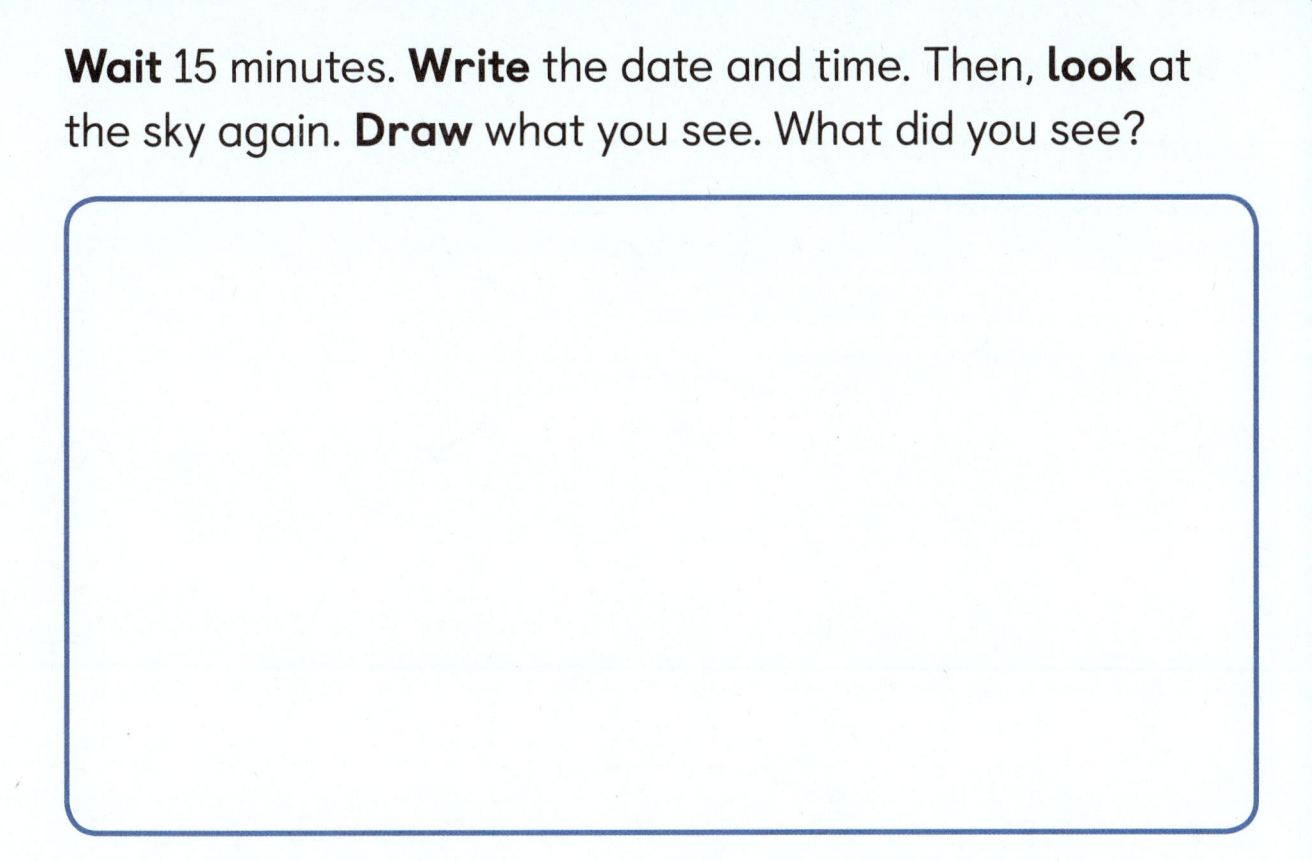

3.4 Moon on the Move

3.4 | Learn How do objects in the night sky move?

Think About the Activity

Look at your drawings. How were they the same? How were they different?

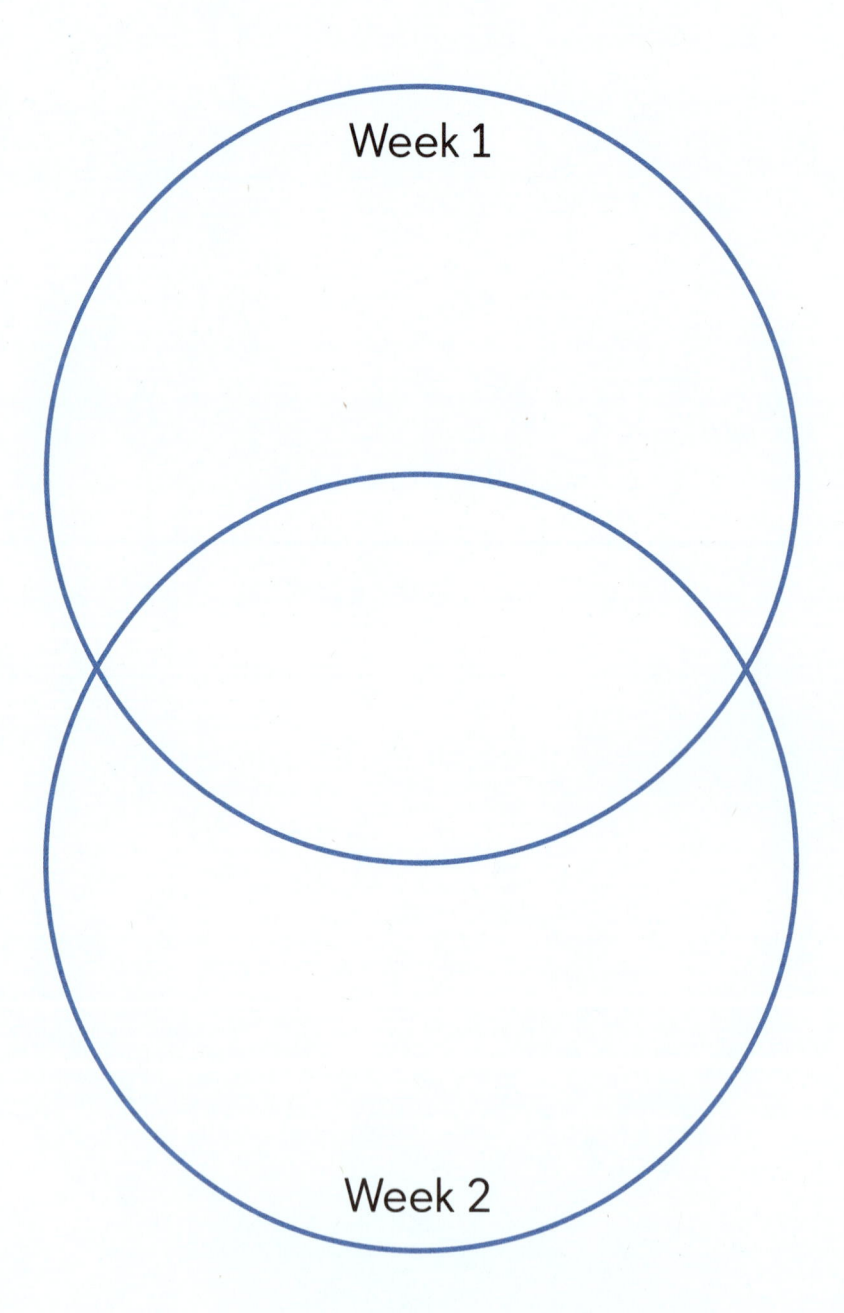

Week 1

Week 2

What patterns did you notice in your observations?

3.4 | Learn — How do objects in the night sky move?

Activity 13
Evaluate Like a Scientist

Quick Code: us1590s

Night Sky

The picture shows how the moon and stars appear tonight.

SEP Analyzing and Interpreting Data

Look at the pictures. **Write** *hour* next to the picture that shows the sky in one hour. **Write** *week* next to the picture that shows the sky in one week.

_____

_____

3.4 Moon on the Move | 187

3.4 | Share How do objects in the night sky move?

Activity 14
Record Evidence Like a Scientist

Quick Code: us1591s

The Night Sky

Now that you have learned about patterns in the night sky, look again at the picture of the night sky. You first saw this in Wonder.

Let's Investigate the Night Sky

SEP Constructing Explanations and Designing Solutions

Look at the Can You Explain? question. You first read this question at the beginning of the lesson.

> 💬 **Can You Explain?**
>
> How do objects in the night sky move?

Now, you will **use** your new ideas about the night sky to answer a question.

1. **Choose** a question. You can use the Can you Explain? question, or one of your own. You wrote questions at the beginning of the lesson.

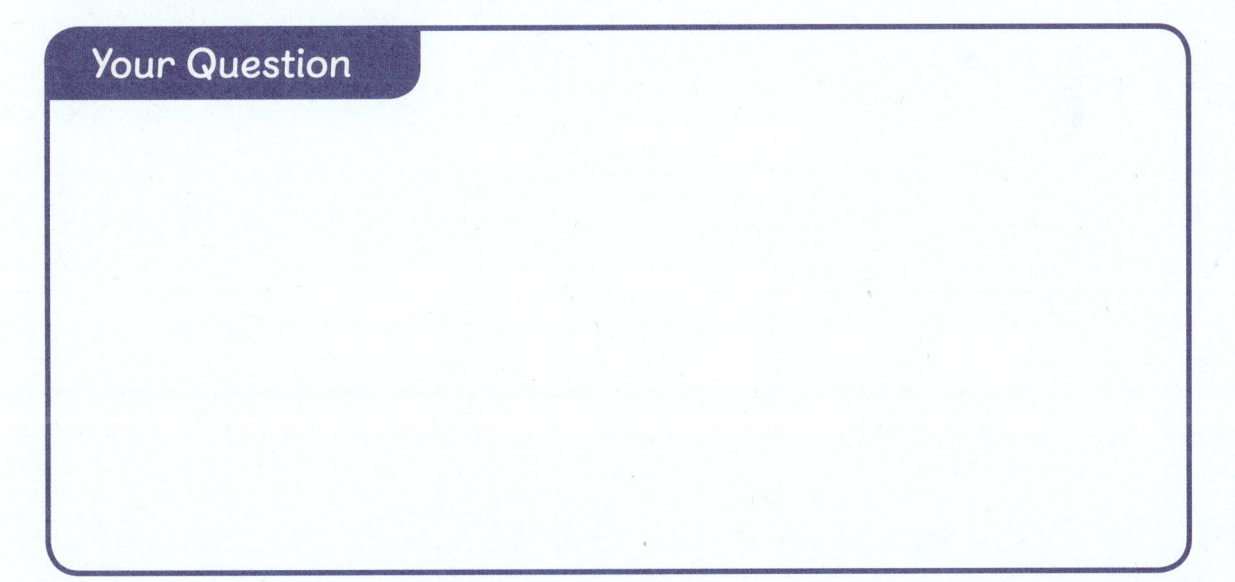

   Your Question

2. Then, **use** the sentence starters on the next page to answer the question.

3.4 Moon on the Move | 189

3.4 | Share   How do objects in the night sky move?

Some objects in the night sky are

_____
_____
_____
_____

The evidence I collected shows that stars

_____
_____
_____
_____

The evidence I collected shows that the moon

_____
_____
_____
_____

Some patterns in the night sky are

_____
_____
_____
_____

STEM in Action

Activity 15
Analyze Like a Scientist

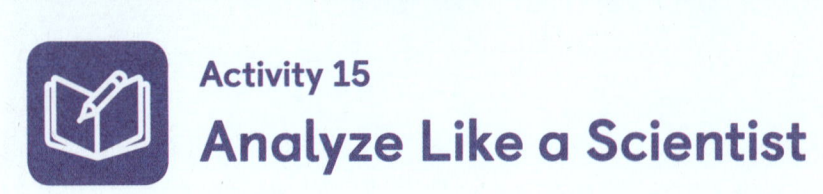

Quick Code: us1592s

Making Lenses

Read the article. **Circle** the sentences that tell about tools we can use to see things that are far away.

Read Together

Making Lenses

It is fun to look at the stars in the sky. People have been looking at the stars for many years. Sailors used the stars to help them know where they were as they traveled across the seas.

SEP Obtaining, Evaluating, and Communicating Information
CCC Scale, Proportion, and Quantity

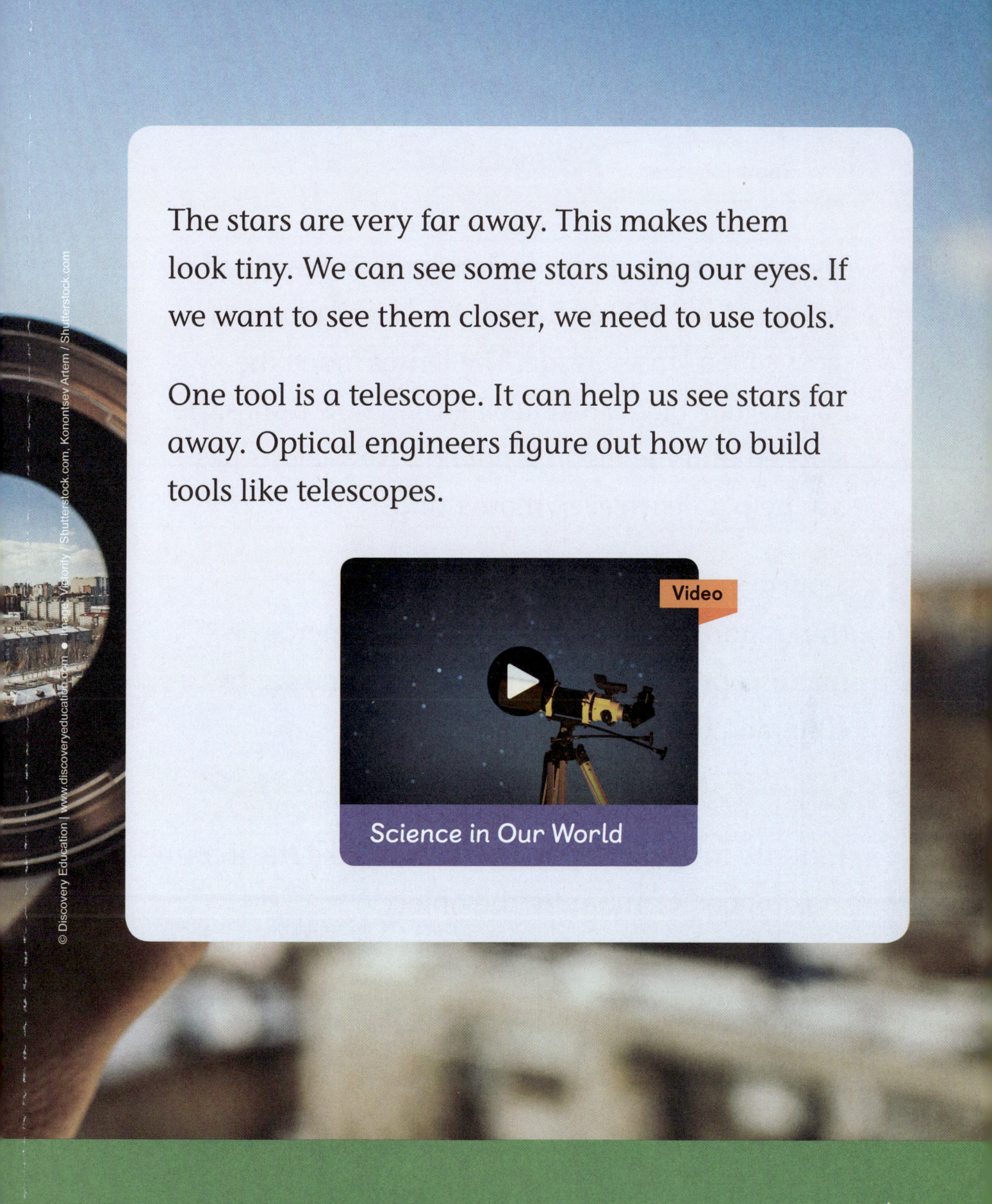

The stars are very far away. This makes them look tiny. We can see some stars using our eyes. If we want to see them closer, we need to use tools.

One tool is a telescope. It can help us see stars far away. Optical engineers figure out how to build tools like telescopes.

Science in Our World (Video)

Read Together

There are many kinds of telescopes. Galileo used one that looked like a long tube. It had pieces of glass called lenses inside. The lenses made the objects look bigger. Galileo learned new things. He learned that the moon is not flat. New telescopes can help us learn even more things!

Optical engineers make lenses for other things, too. The lenses might be in cameras or microscopes. They can change what you see by changing the shape of the glass.

People use binoculars to help them see objects that are far away. The binoculars make the item look bigger, or magnify the object.

Binoculars have a magnification number on them. The number tells how much bigger the binoculars make the object appear. For example, the number 2X means the binoculars make the object look twice as big.

Optical Lenses

Magnification Matching

Draw a line to match each picture with its magnification.

6X binoculars

8X binoculars

10X binoculars

Activity 16

Evaluate Like a Scientist

Quick Code: us1593s

Review: Moon on the Move

Think about what you have read and seen. What did you learn?

Draw what you have learned.

Then, **tell** someone else about what you learned.

Talk Together

Think about what you saw in Get Started. Use your new ideas to discuss why the moon seems to change shape.

SEP Obtaining, Evaluating, and Communicating Information

3.4 Moon on the Move | 197

Unit Project

Design Solutions Like a Scientist

Quick Code: us1594s

Hands-On Engineering: Experimenting with Shadows

In this activity, you will plan a shadow puppet play.

Shadow Puppet Show

SEP Constructing Explanations and Designing Solutions

SEP Asking Questions and Defining Problems

SEP Planning and Carrying Out Investigations

SEP Analyzing and Interpreting Data

CCC Cause and Effect

CCC Patterns

CCC Structure and Function

What materials do you need?
(per group)

- Wax paper
- Cardboard
- Clear plastic lid
- Flashlight
- Batteries, size D

HANDS-ON ENGINEERING

Ask Questions About the Problem

Write or **draw** pictures to show what light sources you need during the day and at night.

Unit Project

What Will You Do?

How many puppets will be in your play?
What materials will you use to make your puppets?
Draw pictures of the puppets for your play.

Build your puppets. **Test** your puppets.
Draw a picture of how you tested your puppets.

Think About the Activity

Write or **draw** your answers to the questions in the chart.
How well did your puppets make shadows during the day and at night?
How could you improve your design?

What Worked?	What Didn't Work?

What Could Work Better?

Grade 1 Resources

- Bubble Map
- Safety in the Science Classroom
- Vocabulary Flash Cards
- Glossary
- Index

Name _____

Bubble Map

Can You Explain?
Question:

Bubble Map | R1

Safety in the Science Classroom

Following common safety practices is the first rule of any laboratory or field scientific investigation.

Dress for Safety

One of the most important steps in a safe investigation is dressing appropriately.

- Splash goggles need to be kept on during the entire investigation.

- Use gloves to protect your hands when handling chemicals or organisms.

- Tie back long hair to prevent it from coming in contact with chemicals or a heat source.

- Wear proper clothing and clothing protection. Roll up long sleeves, and if they are available, wear a lab coat or apron over your clothes. Always wear closed-toe shoes. During field investigations, wear long pants and long sleeves.

Safety Goggles

Be Prepared for Accidents

Even if you are practicing safe behavior during an investigation, accidents can happen. Learn the emergency equipment location in your classroom and how to use it.

- The eye and face wash station can help if a harmful substance or foreign object gets into your eyes or onto your face.

- Fire blankets and fire extinguishers can be used to smother and put out fires in the laboratory. Talk to your teacher about fire safety in the lab. He or she may not want you to directly handle the fire blanket and fire extinguisher. However, you should still know where these items are in case the teacher asks you to retrieve them.

Most importantly, when an accident occurs, immediately alert your teacher and classmates. Do not try to keep the accident a secret or respond to it by yourself. Your teacher and classmates can help you.

Fire Extinguisher

Practice Safe Behavior

There are many ways to stay safe during a scientific investigation. You should always use safe and appropriate behavior before, during, and after your investigation.

- Read all of the steps of the procedure before beginning your investigation. Make sure you understand all the steps. Ask your teacher for help if you do not understand any part of the procedure.

- Gather all your materials and keep your workstation neat and organized. Label any chemicals you are using.

- During the investigation, be sure to follow the steps of the procedure exactly. Use only directions and materials that have been approved by your teacher.

- Eating and drinking are not allowed during an investigation. If asked to observe the odor of a substance, do so using the correct procedure known as wafting, in which you cup your hand over the container holding the substance and gently wave enough air toward your face to make sense of the smell.

- When performing investigations, stay focused on the steps of the procedure and your behavior during the investigation. During investigations, there are many materials and equipment that can cause injuries.

- Treat animals and plants with respect during an investigation.

- After the investigation is over, appropriately dispose of any chemicals or other materials that you have used. Ask your teacher if you are unsure of how to dispose of anything.

- Make sure that you have returned any extra materials and pieces of equipment to the correct storage space.

- Leave your workstation clean and neat. Wash your hands thoroughly.

Vocabulary Flash Cards

axis

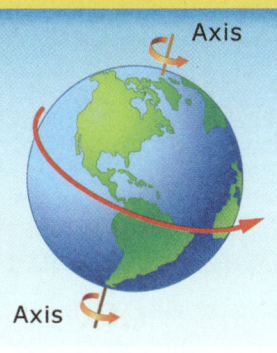

a real or imaginary line through the center of an object; the object turns around it

binoculars

a device that is put up to your eyes so you can see far away

constellation

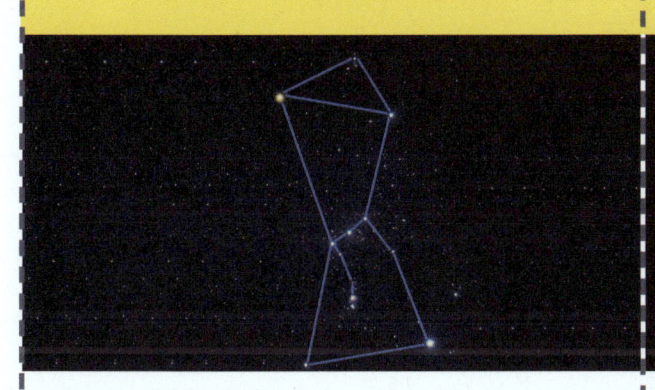

a particular area of the sky; a group of stars

Earth

the third planet from the sun; the planet on which we live

energy
the ability to do work or make something change

engineer
a person who designs something that may be helpful to solve a problem

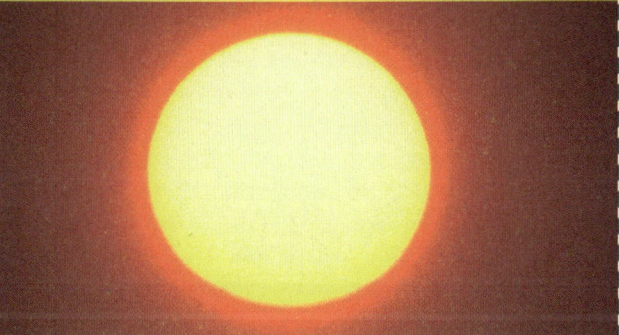

light
a form of energy that makes it possible for your eyes to see

material
things that can be used to build or create something

Vocabulary Flash Cards | R9

moon

any object that goes around a planet

opaque

when no light gets through something, such as wood or metal

orbit

to travel in a circular path around something

position

a place where a person or a thing is located

Vocabulary Flash Cards | R11

reflect

when something like light or heat bounces off a surface

rotate

to turn around a center point; to spin

source

the start or the cause of something

star

a burning ball of gas in space

Vocabulary Flash Cards | R13

sun

any star around which planets revolve

technology

inventions that were developed to solve problems and make things easier

translucent

when some light gets through and what is on the other side might not be very clear, like fog

transparent

when light passes through and you can see clearly, such as clean water and air

Vocabulary Flash Cards | R15

Glossary

English ——— A ——— Español

absorb
to take in or soak up

absorber
tomar o captar

air
an invisible gas that is all around us; living things, such as plants and animals, need it to breathe and grow

aire
gas invisible que nos rodea; todos los seres vivos, como las plantas y los animales, lo necesitan para respirar y crecer

animal
a living thing that moves around to look for food, water, or shelter, but can't make its own food

animal
ser vivo que se mueve para buscar alimento, agua o refugio, pero no puede producir su propio alimento

axis
a real or imaginary line through the center of an object; the object turns around it

eje
línea real o imaginaria que pasa por el centro de un objeto; el objeto gira alrededor de ella

B

binoculars
a device that is put up to your eyes so you can see far away

binoculares
dispositivo que se pone sobre los ojos para poder ver lejos

C

collect
to gather

recolectar
reunir

communicate
to give and get information, messages, or ideas

comunicarse
dar y recibir información, mensajes o ideas

constellation
a particular area of the sky; a group of stars

constelación
área particular del cielo; grupo de estrellas

E

Earth
the third planet from the Sun; the planet on which we live (related words: earthly; earth - meaning soil or dirt)

Tierra
tercer planeta desde el Sol; planeta en el cual vivimos (palabras relacionadas: terrenal; tierra en el sentido de suelo o barro)

edible
able to be eaten as a food

comestible
que se puede comer como alimento

energy
the ability to do work or make something change

energía
habilidad de trabajar o producir un cambio

engineer
a person who designs something that may be helpful to solve a problem

ingeniero
persona que diseña algo que puede ser útil para resolver un problema

Glossary | R19

F

feature
a thing that describes what something looks like; part of something

rasgo
cosa que describe cómo se ve algo; parte de algo

flower
the plant part that blooms with colorful petals and beautiful smells and holds the part of the plant that makes the seeds

flor
parte de la planta que florece con pétalos de colores y aromas agradables y contiene la parte que produce las semillas

fruit
the plant part that contains seeds and grows from a flowering plant

fruta
parte de la planta que contiene semillas y crece de una planta en flor

I

inherit
to receive a characteristic from one's parents

heredar
recibir una característica de los padres de alguien

L

leaf
the part of the plant that grows off the stem and collects sunlight for the plant to make food

hoja
parte de la planta que crece desde el tallo y reúne luz solar para que la planta produzca alimento

light
a form of energy that makes it possible for our eyes to see

luz
forma de energía que hace posible ver con los ojos

M

material
things that can be used to build or create something

material
cosas que se pueden usar para construir o crear algo

measure
to find the amount, the weight, or the size of something

medir
hallar la cantidad, el peso o el tamaño de algo

moon
any object that goes around a planet

luna
cualquier objeto que gira alrededor de un planeta

N

nutrient
something in food that helps people, animals, and plants live and grow

nutriente
algo en los alimentos que ayuda a las personas, los animales y las plantas a vivir y crecer

O

observe
to watch closely

observar
mirar atentamente

opaque
when no light gets through something, like wood or metal

opaco
cuando no pasa luz a través de algo, como la madera o el metal

orbit
to travel in a circular path around something

orbitar
viajar un recorrido circular alrededor de algo

P

plant
a living thing made up of cells that needs water and sunlight to survive

planta
ser vivo formado por células que necesita agua y luz solar para sobrevivir

position
a place where a person or a thing is located

posición
lugar donde se encuentra una persona o cosa

property
a characteristic of something

propiedad
característica de algo

--- R ---

reflect
when something like light or heat bounces off a surface

reflejar
cuando algo como la luz o el calor rebota en una superficie

rotate
to turn around a center point; to spin

rotar
girar alrededor de un punto central; dar vueltas

--- S ---

seed
the small part of a flowering plant that grows into a new plant

semilla
parte pequeña de una planta en flor que se convierte en una nueva planta

seedling

a baby plant that starts from a seed

plántula

planta joven que crece de una semilla

sense

sight, hearing, smell, taste, or touch

sentido

visión, audición, olfato, gusto o tacto

soil

dirt that covers Earth, in which plants can grow and insects can live

suelo

tierra que cubre nuestro planeta en la que pueden crecer plantas y vivir insectos

sound

anything that people or animals can hear with their ears

sonido

todo lo que las personas o los animales pueden oír con los oídos

source

the start or the cause of something

fuente

el comienzo o la causa de algo

star
a burning ball of gas in space

estrella
bola ardiente de gas en el espacio

stem
the part of the plant that grows up from the roots and holds up the leaves and flowers

tallo
parte de la planta que crece hacia arriba desde las raíces y sostiene las hojas y las flores

structure
a part of an organism; the way parts are put together

estructura
parte de un organismo; la forma en que se unen las partes

sun
any star around which planets revolve

sol
toda estrella alrededor de la cual giran los planetas

system
a group of parts that go together to make something work

sistema
grupo de partes que se combinan para hacer que algo funcione

T

technology
inventions that were developed to solve problems and make things easier

tecnología
inventos que se desarrollaron para resolver problemas y hacer más fáciles las cosas

tendril
a long, thin stem that wraps around things as it grows

zarcillo
tallo largo y delgado que se enrosca alrededor de cosas a medida que crece

trait
a characteristic that you get from one of your parents

rasgo
característica que se recibe de uno de los padres

translucent
when some light gets through and what is on the other side might not be very clear, like fog

translúcido
cuando pasa algo de luz y lo que hay del otro lado puede no ser muy transparente, como la niebla

transparent
when light passes through and you can see clearly, such as clean water and air

transparente
cuando pasa la luz y se puede ver con claridad, como el agua limpia y el aire

--- V ---

vibration
the rapid movement of an object back and forth

vibración
rápido movimiento de un objeto adelante y atrás

volume
the loudness of a sound

volumen
la intensidad de un sonido

--- W ---

water
a clear liquid that has no taste or smell

agua
líquido transparente que no tiene sabor ni olor

wave
the way sound moves through the air

onda
manera en la que el sonido viaja por el aire

Index

A

Airline pilots, 154–156
Aluminum, transparent, 103–105
Analyze Like a Scientist, 12–13, 20–21, 28, 29–32, 36–37, 52–55, 64–65, 76, 84–86, 89–90, 103–105, 129, 132–135, 142, 154–156, 173–174, 176, 192–196
Animals
　finding new, 12
　shadows of lizards, 64–65
　in sunlight, 66
Ask Questions Like a Scientist, 10–11, 62–63, 112–113, 162–163
Astronomy, 164–165
Austin, Texas, 156
Axis, 144, 154–155

B

Basketball, light's path after striking, 83
Beam of light
　hitting a target with, 95, 96
　objects that change direction of, 89–90
Binoculars
　lenses for, 194–195
　looking at night sky with, 164–165
　magnification in, 194–196
Blindness, color, 53
Book, light's path after striking, 82

C

Cameras
　lenses of, 83, 194
　pinhole, 41–47
　regular, 41
Can You Explain?, 8, 60, 110, 160
Caves
　exploring, 12–13, 34–35
　seeing in, 10–11, 48–51
　underwater, 10, 12, 15
CDs, 83

Classroom
  adding light to, 19
  light sources in, 19
Clear materials, path of light through, 80
Clocks, 155–156
Clouds, 166, 167
Collecting data, 136–140
Color blindness, 52–53
Colors
  dark and light, 54
  in low light, 55
  seeing, 36, 53
Constellations
  defined, 169
  making, 169–172
  sun and, 174, 175

D

Dark colors, 54
Dark materials, path of light and, 80
Dark rooms
  collecting light in, 36–37
  seeing things in, 11, 18
Data collection, 136–140

Date
  and changes in night sky, 181–187
  and sunrise/sunset time, 137–141
Day
  for airline pilots, 154–156
  cause of, 146
  describing, 114
  Earth from above in, 148–149
  moon in, 176
  rotation of Earth and, 142, 144
  shadow puppet play in, 199
  shadows in, 64
  things in the sky during, 115–117
Daylight, amount of, 136–141
Design Solutions Like a Scientist, 4–5, 198–201
Divers, 11
Doctors, eye, 52–53
Dull surfaces, 97
DVD players, 39

E

Earth
    amount of sunlight on, 110, 151–153
    axis of, 144, 154–155
    day and night from above, 148–149
    distance of stars from, 29, 30, 173
    distance of sun vs. moon from, 166, 167
    rotation of, 142, 145–147
    source of light on, 31
East, sunrise in, 132, 133
Energy, light, 20, 87
Engineers
    material design by, 104
    optical, 193, 194
Evaluate Like a Scientist, 16–17, 27, 47, 56, 68–69, 88, 98–99, 106, 115–117, 130–131, 148–149, 157, 166–167, 175, 186–187, 197

Exploration
    of caves, 12–13, 34–35
    of outer space, 14
    of shadows, 126–127
Eye doctors, 52–53
Eyeglasses, 53
Eye(s)
    collecting light in, 36
    fixing color blindness in, 53
    reflection of light to, 90, 97
    seeing stars with, 193

F

Flashlights
    exploring with, 12
    path of light from, 23–26
Frosted materials, 80

G

Galileo, 194
Gases, in stars, 29, 173
Glass
    properties of, 105
    transparent, 103–104

H

Heat, 82–83
Houses, light inside and outside, 70

I

Image, from pinhole camera, 43–45
Investigate Like a Scientist, 23–26, 41–46, 71–75, 78–81, 92–96, 119–123, 180–185

J

Jupiter, 167

L

Lenses
   of camera, 83, 194
   making, 192–196
Light. *See also* Path of light; Reflection of light
   in action, 68–69
   adding, to classroom, 19
   after hitting an object, 60, 76, 101
   angle of, 126–127
   beam of, 89–90, 95, 96
   changing, for a job, 52–53
   colors in low, 55
   defined, 22
   discussing, 17
   heat, sound, and, 82–83
   movement of, 20–21, 23–26, 47, 50, 51
   objects that change direction of, 89–90, 102
   passing of, through materials, 75, 78–81, 84–86, 102
   position of, 87
   scavenger hunt for, 18–19
   and seeing objects, 8, 48–49
   shape of, 27
   shining, on shadow puppets, 73–75
   and stars, 29–32
   transmission of, 69, 74–75
   using, 36–40

Light bulbs, paths of light from, 23–26
Light colors, 54
Light energy, 20, 87
Light rays, from sun, 30
Light sources
    in classroom, 19
    discussing, 56
    identifying, 17, 19, 50
    made vs. natural, 17
    and movement of light, 20–21
    in night sky, 28
    path of light from, 23–26, 50, 51
    for pinhole camera, 46
    pretending to be, 33–35
    for shadow puppet show, 4–5, 199
    in underwater caves, 10, 15
    what you know about, 16
Lizards, shadows of, 64–65
Location, amount of sunlight and, 132–135
London, England, 134–135
Los Angeles, California, 156

M

Magnification, 194–196
Magnification number, 195
Materials
    frosted, 80
    passage of light through, 75, 78–81, 84–86, 102
    and path of light, 106
    reflection of light by, 92–96
    transparent, 103–104
Metals, 85, 104
Microscopes, 194
Midnight, 145
Mirrors
    path of light after striking, 82, 83
    reflecting light with, 89–90
    scratched, 91
Models, 172

Moon
- binoculars and telescope for looking at, 165
- cycles of, 178–179
- date/time and changes in, 180–187
- describing, 150–153
- distance from Earth to sun vs., 166, 167
- light from, 28, 51
- movement of, 166–167, 191, 197
- phases of, 166
- seeing, during day, 176
- stars vs., 32
- and sun, 112–113

N

Nature, light sources in, 17

Night
- for airline pilots, 154–156
- cause of, 146
- describing, 114
- Earth from above during, 148–149
- Earth's rotation and, 142, 144
- shadow puppet play at, 199
- shadows and, 64
- things in the sky during, 115–117

Night lights, 40

Night sky
- binoculars and telescope for looking at, 164–165
- cycles in, 178–179
- investigating, 162–163, 188
- lenses for looking at, 192–196
- movement of objects in, 160, 189
- objects in, 160, 163, 166–168, 175
- observing changes in, 180–185
- patterns in, 169–172, 185, 191
- sources of light in, 28
- stars in, 29–30

Noon
    midnight and, 145
    shadow length at, 130, 131

O

Object(s)
    behavior of light after
        hitting, 60, 76, 101
    changing direction of
        light with, 89–90, 102
    and image from pinhole
        camera, 43–45
    light for seeing, 8, 48–49
    light shining through, 68,
        75, 102
    movement of, 69
    in night sky, 160, 163,
        166–168, 175, 189, 190
    opaque, 76, 85, 86, 88, 102
    seeing through, 84–85
    translucent, 76, 85, 86, 88
    transmission vs.
        reflection of light by, 74
    transparent, 76, 84, 86, 88

Observe Like a Scientist, 14,
    15, 18–19, 22, 33–35,
    38, 39–40, 66, 67, 70,
    82–83, 87, 91, 97, 114,
    118, 124–125, 126–128,
    143–145, 146–147,
    164–165, 168, 177,
    178–179
Ocho Rios, Jamaica, 134–135
Opaque objects
    ability of light to travel
        through, 85, 86
    effect on light of, 76, 102
    identifying, 88
Optical engineers, 193, 194
Optometrists, 52–53
Orbit, 176
Outer space, exploring, 14

P

Palau, 15
Paper
    path of light after
        striking, 80, 83
    as translucent material, 85

Path of light
   light source and, 20–21, 23–26, 50, 51
   materials and, 106
   from night light, 40
   in pinhole camera, 47
   and reflection, 98–99
Patterns
   in night sky, 169–172, 185, 191
   in shadows, 123, 125, 127
   of stars, 169–172
   of sun, 179
   of sunlight, 118, 123, 153
   in sunrise and sunset time, 141
People, light sources made by, 17
Phases of moon, 166
Pilots, airline, 154–156
Pinhole cameras, 41–47
   light traveling in, 47
   making, 41–42
   using, 43–47
Plants, finding new, 12
Plastic, transparent, 104

Position, of sun, 123
Puppets, shadow. *See* Shadow puppets

R

Rainbows, 40
Record Evidence Like a Scientist, 48–51, 100–102, 150–153, 188–191
Reflection of light
   changing direction with, 89–90
   from dull and shiny surfaces, 97
   investigating materials ability to reflect, 92–96
   by moon, 28
   transmission of light vs., 69, 74–75
Rotation of Earth
   as cause of day and night, 146–147
   direction of, 145
   and movement of sun in sky, 142

S

Safety vests, 55
Sailors, stars for, 192
Sanghera, Jas, 104
Scratched mirrors, 91
Seattle, Washington, 134–135
Seeing
    in caves, 10–11, 48–51
    collecting light and, 36–37
    in dark rooms, 11, 18
    light for, 8, 48–49
    shadows, 63
    through objects, 84–85
Shadow puppets
    light source for, 4–5
    making, 71–72
    planning a play with, 198–201
    shining light on, 73–75
    testing, 200–201
    theater for, 2–3
Shadows
    and amount of light, 64–65
    and behavior of light, 62–63
    changes in, 67, 119–123, 126–127
    describing, 100–101
    experimenting with, 4–5, 198–201
    exploring, 126–127
    measuring, 119–123
    patterns in, 123, 125, 127
    playing with, 67
    and position of light, 87
    seeing, 63
    and sunlight, 66
    time of day and length of, 129–131
Shadow stick, 124–125
Shadow theater, 2–3
Shell Beach, 118
Shiny surfaces, 97
Shutters, 70
Skits
    about exploring caves, 34–35
    about rotation of Earth, 146–147

Index | R37

Sky. *See also* Night sky
    cycles in, 143–145
    movement of sun across,
        119, 124–125, 128, 142
    things you can *see* in,
        115–117
Sound, 82–83
Space, exploring, 14
Spinel, 104, 105
Stars, 166
    date/time and changes
        in, 180–187
    during day, 176
    investigating, 190
    lenses for looking at,
        192–196
    and light, 29–32
    moon vs., 32
    movement of, 177
    patterns of, 169–172
    sun vs. other, 173–174
Sun
    and constellations, 174, 175
    describing, 150–153
    distance from Earth to
        moon vs., 166, 167
    light from, 28–31, 51
    light rays from, 30
    and moon, 112–113
    movement of, across sky,
        119, 124–125, 128, 142
    other stars vs., 173–174
    patterns of, 179
    position of, 123
Sunlight, 157
    amount of, on Earth, 110,
        151–153
    location and amount of,
        132–135
    patterns of, 118, 123, 153
    and shadows, 66, 128
    what you know about,
        115–117
Sunrise
    defined, 133
    shadow length at,
        128–129, 130–131
Sunrise time
    collecting data on,
        137–140
    and location, 134–135
    patterns in, 141

Sunset
    defined, 133
    shadow length at, 128, 130, 131
Sunset time
    collecting data on, 137–140
    and location, 134–135
    patterns in, 141
Surfaces, reflection of light from, 97

T

Telescopes
    learning about space with, 14
    lenses for, 193–194
    looking at night sky with, 164–165
Theater, shadow, 2–3
Think Like a Scientist, 136–141, 169–172
Time
    for airline pilots, 154–156
    changes in night sky over, 181–187
    shadow length and, 129–131
    sunrise, 134–135, 137–141
    sunset, 134–135, 137–141
Translucent objects
    ability of light to travel through, 85, 86
    effect of light on, 76
    identifying, 88
Transmission of light, 69, 74–75
Transparent objects
    ability of light to travel through, 84, 86
    aluminum, 103–105
    effect of light on, 76
    identifying, 88

U

Underwater caves, 10, 12, 15

W

Water, light and, 83
West, sunset in, 132, 133
White materials, path of light and, 80

Window(s)
  light traveling through, 84
  path of light after striking, 82
  with shutters, 70

Y

Year, changes in sunlight over, 134–135